THE TRUE ESSENCE OF VACUUM

Reality is not what you think it is

BOGDANOV

The True Essence of Vacuum

Reality is not what you think it is

© Copyright 2020 George Warehouseman
All Rights Reserved.
Protected with www.protectmywork.com,
Reference Number: 9178310720S030

(Book Formatting by Derek Murphy @Creativindie)

ACKNOWLEDGEMENTS

In short, I have nobody to thank for this book, really! However, there are some people and organizations whose actions contributed to the creation of this book. But that wasn't a contribution as you might be imagining it. Basically, it was the ignorance and blindness of some science journals and scientists towards clear and strong arguments and facts that led to the creation of this book.

In particular, I would like to thank the science journal named "**Annals of Physics**" and their editor in chief along with the whole Editorial Office. It is thanks to their refusal even to look into this topic let alone to consider the arguments and publish the hypothesis based on them that resulted in this book.

This science journal confirmed what **John Wheeler** once said. He said this: *Science journals reject to publish your ideas and arguments not because they don't understand them, but because they do. If they don't understand something, they publish it.*

After all, over years science journals have published a large number of most bizarre ideas which nobody can understand or test. And here you have one hypothesis that is based on facts, makes perfectly logical sense, has suggestions for experiments and makes predictions that are impossible within the existing theories, would deliver a huge contribution to our understanding of Reality, and it got blocked by the bureaucratic machine of science officials.

CONTENTS

INTRODUCTION 1

CHAPTER ONE
About how this book is organized 3

CHAPTER TWO
The Description of existing understanding of Vacuum 5

CHAPTER THREE
The short description of what is Vacuum and how it works in my hypothesis 9

CHAPTER FOUR
The enormous importance of the True Essence of Vacuum in Cosmology and in Science in general 15

CHAPTER FIVE
The full description of what is Vacuum in my hypothesis, suggestions for experiments to test it, and the predictions following these tests 27

CHAPTER SIX
The Blunders of Time Dilation 67

CHAPTER SEVEN
The images depicting the True Pattern of Magnetic Field 77

INTRODUCTION

The True Essence of Vacuum has a huge importance in science on multiple levels and in multiple fields. And yes, I strongly believe that presently we have got it wrong. The biggest problem is the lack of ideas of how to experimentally test the True Essence of Vacuum. And this is exactly what I intend to change with this book.

After having read this book, you will be very knowledgeable about the Vacuum and topics it affects in science. You will be surprised how many and how important topics and even whole fields of science, especially in Theoretical Physics, depend on what we think about Vacuum. For this reason, I confidently declare that Vacuum is so far the absolutely number one topic that can turn around nearly everything in Theoretical Physics. This will affect the Nature and Essence of Light, Expansion of the Universe, Dark Energy, Dark Matter, Big Bang Theory, and the list goes on and on. This question about the True Essence of Vacuum will help to explain Black Holes, Magnetism and Gravity itself, and many, many other things we are presently struggling with.

Later in the text I will make a list, with short description to them, of the topics that Vacuum affects. Trust me, after having read this book, you will have enough tools to challenge not only your own perception of the True Essence of Vacuum and Reality itself, but you will have arguments and facts to challenge all your physics teachers and scientists of all levels.

The question right now is this – Are you curious and do you want to understand the world around you? If the answer to this question is yes, then challenge yourself by jumping into this interesting and captivating story about the True Essence of Vacuum! And no, the existing theorise are absolutely wrong about it. I can and will prove it with experimental facts in this very book.

Also, at the end of this book, I will show to you what dark secrets the theory of Time Dilation is withholding from you. For those who are interested in Magnetism, I have prepared some images that describe the True Pattern of Magnetic Field. Sorry, there will be no further descriptions about how it works in this book. But you can have some short insights of it if you check my other book – **New Theory of Where We Come from and How the Universe Works**. But the full story about Magnetism and Gravity is on the way. And the correct understanding about the Vacuum is absolutely essential to explain it.

So, happy reading!

CHAPTER ONE

I have said this many, many times already, but this is very important, and, because of that, I will repeat it again – the True Essence of Vacuum is extremely important in understanding large number of things in science, and first of all, many things in Cosmology. For this reason I decided to publish here my whole manuscript about the Vacuum which I have sent to science journals. Needless to say that it was rejected.

I hope that ordinary people are far more open minded than scientists are. So, happy reading!

About how this book is organized!

Dear Reader,

I hope that after having read this paper you will change your views completely on what is Vacuum and what is its true nature. Also, I hope that you will be very interested in results of experiments that I am suggesting in this book in order to test my hypothesis of what is the True Essence of Vacuum.

The structure of this book will be as follows:
1. Description of the existing understanding of what is Vacuum.
2. The short description of what is Vacuum and how it works in my hypothesis.
3. The enormous importance of the True Essence of Vacuum in Cosmology and in Science in general.
4. The full description of that is Vacuum in my hypothesis, suggestions for experiments to test it, and the predictions following these tests.

It is important to keep in mind that the predictions of the tests suggested are absolutely impossible within the existing understanding of what is Vacuum.

THE TRUE ESSENCE OF VACUUM

CHAPTER TWO

The Description of existing understanding of Vacuum.

Oxford Learners Dictionary on **what is Vacuum**:

- *a space that is completely empty of all substances, including all air or other gas*

The next quote is borrowed from a website titled:

https://plato.stanford.edu/entries/newton-stm/

2. The Legacy from Antiquity

2.1 The Void

*The most important question shaping 17th-century views on the nature of space, time and motion is **whether or not a true void or vacuum is possible**, i.e., a place devoid of body of any sort (including rarefied substances such as air). Ancient atomism, dating back at least to the pre-Socratic philosopher Democritus (5th century, B. C.), held that not only is such possible, but in fact actually exists among the interstices of the smallest, indivisible parts of matter and extends without bound infinitely. Following Plato, **Aristotle rejected the possibility of a void, claiming that, by definition, a void is nothing, and what is nothing cannot exist.***

As you can see, over time the views of what is Vacuum have changed. Right now we have stopped on the assumption quoted previously as described by Oxford Dictionary. Namely, currently our science thinks that Vacuum is a completely empty space in which there is no Matter of any sorts at all or close to that. And even though there might be in a Vacuum some Particles of some substances, any of such Particles still would not be connected among them.

I have to admit that I fully side with Aristotle on this - **a void is nothing, and what is nothing cannot exist...**

THE TRUE ESSENCE OF VACUUM

Our present assumption about what the Vacuum consists of and how it works derives mainly from five facts. Slightly later, as strange as it sounds, I will present these facts to you. Interestingly, it might be the only list where all of these facts, which have led to conclusions of our science about what the True Essence of Vacuum is, are nicely listed in one place. Regardless of me being right or wrong with this claim, I have to say that I have never seen anyone ever addressing the problem of Vacuum in a scientific way, let alone carrying out any experiments to prove or disprove the existing claims. And this is true with one of the most important (iconic?) questions in Cosmology and Theoretical Physics. Later I will explain to you what these topics are that this question has been affecting and how it has paralyzed our advances in understanding of the Universe.

Of course, some experiments have been carried out to prove the observable facts, but not what causes them, let alone what is the True Essence of Vacuum. From these experiments we have learned some facts about the behaviour of the Vacuum, but that doesn't explain its True Essence. To understand this problem think about the Gravity. The fact that we can measure and predict the effects of Gravity does not explain why the Gravity works in this particular way and what causes it. So, we still don't know what Gravity is. We only know how it behaves.

And the same is true with respect to the Vacuum. For this reason, in this book, I will try to explain the Essence of Vacuum in a different way and I will try to prove my hypothesis to you with some experiments that I have already carried out by myself. Also, I will provide some suggestions for further experiments that I cannot carry out by myself. And the predictions of those experiments are such that are absolutely impossible within the existing theories.

So, next comes the list of arguments that had led our science to the existing claim about Vacuum and Space both being Particle-free areas or areas where there is no Matter in them.

A list of facts that our science cannot explain, and which led to the claim that Vacuum is *"a space that is completely empty of all substances, including all air or other gas"*, is as follows:

1. **The Sound does not propagate in the Vacuum**, therefore there are no Particles, a.k.a. - there are no physical substances in the Vacuum as the Sound needs a medium consisting of uninterrupted chain of Particles to be able to propagate forward. And even if there were some Particles, they are still not connected among them and that stops the Sound there.

2. **You cannot use pumps in the Vacuum because pumping doesn't work there**. In essence, if you tried to pump a balloon in the Vacuum with a conventional pump, you would fail to achieve anything. For this reason, there cannot be any Particles in the Vacuum.

3. **Substances easily break apart**. For instance, you can take a piece of Water and split it in as many separate parts as you desire. So, clearly, Particles of substances do not hold together in the same way as Particles do in solid substances. But that can only lead to one conclusion – whenever the given substance, or combination of substances, is very rarefied (scattered), there the Particles of these substances must be unconnected. And even more unconnected they must be in the Vacuum for exactly the same reason even if in the Vacuum there happened to be some Particles.

4. **When we are moving through substances, such as Air or Water, It feels as if the Particles of them are easily giving up and separating from each other**. It definitely feels like that intuitively. After all, if that weren't true, and the Particles of Liquid Substances were holding together with each other instead, we wouldn't be able to move through them. This is how it feels intuitively. Therefore we assume that Particles of substances never hold

together, even though we don't have any experimental proof for that. Everything in this claim is based on our intuitive feeling.

5. **All objects in Vacuum fall at the same rate**. This is also clearly true, as many experiments have confirmed it. And this is being used as one of the strongest arguments to support the claim that Vacuum is Particle – free area. I have to admit that all these are absolutely sane claims, as the density of Particles of a given substance indeed affects the falling rates of objects. So, if there are no Particles, then nothing would slow down the falling object, right!? So, the emptiness is proved, and the case is closed, right!? Not if this fact, and all the others, can be explained in a different way and still make a perfect and logical sense. If there is a theoretically possible way to explain the Essence of Vacuum that would also explain all observable facts and experimental results, then we should consider that another hypothesis too. At least scientific method would demand us to do that.

All these facts (above in **Bold** and underlined) are undeniably true. Also, seemingly true appear to be conclusions made from these facts. After all, those are absolutely logical conclusions. However, in this book I am going to explain how these observations (facts) can be explained within the Space/Vacuum that is fully filled with substance whose Particles are in close and permanent contact with each other and with all other objects that they touch.

The explanations of these facts in my hypothesis you will find randomly located throughout this book. The only things that would be important in this story would be the suggested experiments and predictions that should be unique and possible only if my hypothesis is correct. And I have just such suggestions.

CHAPTER THREE

The short description of what is Vacuum and how it works in my hypothesis.

My hypothesis is based on a few other assumptions that contradict the existing ideas about the True Nature of Smallest Particles and properties of Liquid Substances in general. It would be very helpful for the reader to keep their mind open up to the very end of this book, as only at the end all pieces of this puzzle will form into a completed and logical image. After all, Theoretical Physics are being called "theoretical" for a reason. This means, many of ideas describing reality in Theoretical Physics are about entities and their properties that are too small or too large to be measured. This is why we develop ideas simply based on assumptions about what could be possible and what could be the properties of some individual players in these stories and what could be their mechanics. When the hypothesis is fully formed, we suggest ideas for experiments to test them and make predictions about what would happen if we are right about these assumptions. And whichever suggestion survives the tests and predictions are declared to be the best theories to date and end up being taught at schools. So, yes, most ideas in Theoretical Physics, like the sizes, nature, and the very existence of Electrons (and all other subatomic Particles for this matter), the Big Bang Theory, existence of the Space-Time Fabric etc., are only theoretical. Some of these things are even impossible to measure and prove experimentally directly.

In case of my hypothesis, a completely new idea of Mechanics of Smallest Particles was necessary. I cannot suggest any tests to prove these mechanics right now, but I can and will suggest experiments to test and prove certain consequences caused by them. Those will be tests with predictions that are only possible if my hypothesis is correct and will be impossible withing the existing theories. All relevant suggestions for experiments and predictions that should follow I will reveal in due time.

So, all Smallest Particles in my hypothesis are like tiny Magnets. They have Poles and they have Attraction and Repulsion qualities to them similarly to how it is with Magnets. Also, all Smallest Particles are rotating. They rotate in a very specific direction with respect to their Poles. In this they are similar with Cosmic Objects.

Also, if you squeeze these Particles, they push back. If you stretch them, they pull back as individual entities and also as one united body as all Particles of all Matter are connected among them. Keep in mind that all Particles in my hypothesis hold together at all times.

I know that it is very challenging to comprehend and accept this right now. For this reason, I urge you to keep in mind that I will suggest experiments and provide the predictions that will prove me right if I am correct.

So, finally, what does Vacuum consist of in my hypothesis, right!? In essence, there is no such thing as an empty Space where "empty" would mean a Particle-free area (no Matter in it) anywhere. The reason for that is, as Aristotle has put it rightly, as follows – anything that is free from Matter is as good as things that doesn't exist. Existence itself is determined by the existence and presence of the Matter. If there is no Matter anywhere, so this place doesn't even exist. Namely, a place where there is no Matter doesn't even exist. And for this reason, there cannot be such thing as independently flying around Particles which aren't connected among themselves. In my hypothesis such state of empty (non-existent) places is impossible. Instead, all Particles of all substances are holding together to each other at all times in the same way as do Particles of solid substances. What makes them appear to be fluid is their ability to roll over each other from one connection to others. Also, these Liquid Substances are attached to all objects they touch, and they roll over them in the same way as they would roll over each other.

How does the movement of Particles in Liquid Substances work?

Even though Particles of Liquid Substances keep together at all times and very strongly, they also keep rotating and can move from the engagement at a given moment with given Particles to others by rolling over to them and without ever losing the grasp with the nearest objects/Particles to them. I understand that it feels counter-intuitive, but there are experiments that would produce results which can only be possible if this hypothesis of mine is correct.

How would these Particles roll over each other? Imagine as if they are like tiny cogwheels that are connected in one system where the movement of one cogwheel would cause the movement of others to which it is directly connected. And they would stay closely and directly connected with some Particles at all times. The only differences between cogwheels and Particle behaviour in my hypothesis would be that cogwheels aren't holding to each other, but the Smallest Particles in my hypothesis would be. Also, the rotational speed transferred from Particle to Particle at certain distances would gradually decrease as there would be a certain resistance to any impact on Particles that forces them to behave differently compared to what the surrounding environment around them has already. And that too wouldn't be the case with cogwheels.

For the reason that all Particles are holding together at all times, and I am repeating it because it is very important, when you would try to squeeze any substance, it (every single Particle of it as a united system) would increasingly push back. And when you would try to expand any substance, it (every single Particle of it as a united system) would increasingly pull back. And there would be no force possible that would actually break them apart as the resistance in both cases would grow extremely rapidly following the increase of the impact, be it squeezing or stretching… This, by the way, is the short version of what is the True Essence of Vacuum. No worries, the proofs and deeper explanations will follow in due time.

So, all in all, I hope that I have delivered the short version of the mechanics of Smallest Particles and Liquid Substances as they appear in my hypothesis. Now let us take a look at what problems in contemporary Science, especially in Cosmology and Theoretical Physics, my hypothesis would help to solve.

As for the lack of sound and pumps not working in the Vacuum, which were two most important facts that led to interpretation of the Vacuum as a Matter-free area, this is perfectly explainable with my hypothesis too. You will see in this book how easy and logically simply you can explain the lack of sound and conventional pumps failing in a Vacuum despite it, according to my hypothesis, is fully filled with Particles of a Substance. And these Particles would be fully connected with each other and with the surrounding objects (and walls) that they can touch at all times. Trust me, at the end of this book your present understanding of the Reality will have cracks in it.

THE TRUE ESSENCE OF VACUUM

CHAPTER FOUR

The enormous importance of the True Essence of Vacuum in Cosmology and in Science in general.

I found this topic absolutely fascinating and of enormous importance to our understanding of how the Nature works. Unfortunately, I didn't find any serious research dedicated to the exploration of this question ever, let alone any experiments conducted in order to understand it. And this very sad as it sometimes appears that scientists aren't even interested in this topic, as if they are not interested in understanding the true mechanisms behind the Physical Reality around us to start with.

We live in times when technologies allow us to do mesmerizing things. We have built the Large Hydron Collider that costed some 7 billion and we are spending around 1 billion per year to keep it running. And yet, there are some very important topics in science, which would affect large number of our assumptions about the Reality and how it works in Theoretical and Applied Physics, which would be much cheaper to explore and yet, we have been ignoring them completely. This is especially bizarre because explorations of some of these topics, like about the Vacuum of Redshift (Dopplers Shift itself…), would cost us far less and provide far greater benefits than the LHC will ever do. And nobody seems to be even interested in that… Yes, the Doppler's Shift needs serious revisiting along with the Redshift. And my next book is going to be exactly about that…

So, what fields of science would be affected if the Vacuum were as it is described in my hypothesis? I will give this list to you very soon. After having provided the list, which will include only the most important topics that would be affected, I will provide a short description of how exactly each of these questions will be affected. And, when that is finished, I will explain my hypothesis about 'What is the True Essence of Vacuum" in full.

The list of topics that the Essence of Vacuum (as it is in my hypothesis) will help to explain better and in a different way is as follows:

1. The Essence of Light.
2. The Nature of the Speed of Light.
3. The **Redshift** and the so-called **Expansion** of the Universe, **Big Bang** theory, **Dark Energy**.
4. The Black Holes.
5. The Double Slit Experiment.
6. The Photoelectric Effect.
7. Why all dependent Cosmic Objects are floating in one plane around their host Cosmic Object and this plane happened to be around the Equator of the host.
8. How all Electromagnetic Phenomena propagate.
9. Magnetism and Gravity.
10. The Dark Matter.

As promised, next follows the more detailed description of the affected topics in science because of the different understanding of the Vacuum. This part of the story you can skip for now and jump straight to the full description of my hypothesis of the True Essence of Vacuum. But if you are interested in what it could change, then keep reading.

1. **<u>The Essence of Light.</u>**

    As we know it, Light behaves like a wave. I would actually say that it creates waves... But since we assumed that Space is a Matter-free area, we were faced with a dilemma about what creates these waves and in what Substance, as waves need one, and how does Light propagate in Space if there is no Substances up there at all. Also, we couldn't explain what Light is because of this assumption of ours that Vacuum is an empty space. This led to the idea of the new Particle called Photon, where Photons are Particles that have nothing to do with conventional Particles whatsoever. In fact, Photons have properties that are impossible to prove and measure, since they have no Mass and they don't slow down ever. In reality, these even aren't properties of Particles at all... But, if there is a Substance in Space that fully fills it in the same way as oceans are filled with Water, then we can find other interpretations and explanations that would explain the True Essence and Nature of Light in a completely different way. In my hypothesis, the Light is a certain Charge of Particle that propagates in direction from the Source of Light by jumping in a copy/paste manner from Particle to Particle. Clearly, this interpretation needs the Space being filled with a Substance which Particles would be tightly and permanently connected to each other. In this way we could explain absolutely everything with respect to the observable facts about Light. But this is a story for another time.

2. **The Nature of the Speed of Light.**
Clearly, the Speed of Light is part of the Essence of Light. However, they can easily be separated as different topics as each of them are differently important due to their application. So, if the Light needs a medium to propagate, then it is dependent on this medium. But that means, its speed might be affected by the movement of this medium itself. This is exactly how it is in case of another Electromagnetic Phenomenon – Electricity. And that would mean the end of the era of assumption that Speed of Light is constant with any frame of reference. And, indeed, these differences in the Speed of Light would be measurable and provable. Be aware that I have already a few ideas of how to test that.

3. The **Redshift** and the so-called **Expansion** of the Universe, **Big Bang** theory, **Dark Energy**.
You might be wondering why all these topics are under one subtitle. The thing is - they all derive from the same

source – our assumptions of what is Redshift and what causes it. So, if my hypothesis is right, not only it would explain the wave - function of Light, but it also would perfectly explain the causes of the Redshift in a completely different and completely normal way. Namely, when waves are moving forward from the source of them, they gradually flatten and stretch out over distances. This means, if the Light is an effect propagating through a medium of whatever Substance has filled the Space, then a mere distance away from the Source of Light would affect how stretched these waves would appear to us. So, the further the observed object would be, the more "redshifted" their Light would appear to be. And that would annihilate at a stroke the claim that Redshift proves the Expansion of the Universe, let alone Expansion that accelerates. Accelerating Universe, as it is presented in our textbooks, means that the Earth is indeed at the center of the Universe from which everything moves away. After all, if the objects that are further accelerate faster, then this is no longer equally expanding Universe anymore. Then the Earth simply must be at the center of the Universe. And that is extremely bizarre thing to think about, unless something is wrong with our perception of the Reality. But if my hypothesis is correct, then more Redshift from more distant objects is exactly what we should expect and there would be no need for interpretations that everything is moving away from each other in Space That would mean - problem solved… And based on this all problems, such as the Big Bang theory, Expansion of the Universe and Dark Energy would instantly cease to exist.

4. **Black Holes** will be no longer a mystery.
   Black Holes are a mystery precisely because we have misunderstood and misinterpreted the True Essence of Light. But Light cannot be explained properly if we live

with this assumption that Vacuum is Matter – free area. So, clearly, the True Essence of Vacuum is absolutely important in understanding Black Holes too. After all, if the Space is filled with the Substance, and Light/Information propagates through this Substance as it is a medium for it, then the movement of this medium would affect what and how we see distant objects. Just think about the fact that the so – called Black Holes are extremely fast rotating objects. If that is true, then, combined with my idea of how the Light propagates, it becomes easy to explain the "Black Hole phenomenon" within the existing Laws of Physics. In my hypothesis the rotational speed of those objects simply distorts the Substance around it so much that any Information coming from these objects appears to us black in the same way as the Space appears black if you are looking sideways from the Sun even when the Sun shines straight at you.

5. **The Double Slit Experiment.**

First of all, if Vacuum as Matter – free area doesn't exist, then the Energy of Light is propagating through Particles of the given Substance and, while doing that, is causing waves. So, the wave function itself can be perfectly explained in this way. Also, observing the Light, if it works as in my hypothesis, by applying another Source of Light would directly affect the very environment through which the original Light of this experiment was moving through. Clearly, that other Source of Light had to affect the results of the experiment. This way of "observing" the behaviour of Light by using another Source of Light is as wrong as to test the strength of Electrical Charge in a wire by applying another Electrical Charge to the same wire and ignoring that both effects would combine and change the results altogether... But since scientists have got this all completely wrong, they cannot understand the results of the Double Slit Experiment ever since it delivered those results…

6. **The Photoelectric Effect.**
   Clearly, Photoelectric Effect is easy explainable if the Light is and works as described by my hypothesis, for which we need to prove that Vacuum is area with stronger or weaker stretched Particles of a given Substance in a given area. In my hypothesis, as I have said it many times previously already, Photoelectric Effect would be caused by a level of charge that moves from Particle to Particle in a copy/paste manner, similarly to how Temperatures propagate. The only difference between the two would be the speed of propagation of Light that is much faster. So, since the Light is a certain level of Energetical Charge of Particles, it can cause this Charge to be transferred to certain metals if a certain strength of Energy of Light has been reached. Clearly, to break the resistance of Particles of metals, the level of Charge of Light must have a certain strength. As simple as that. What about the different colours of the Light where some colours never trigger Electricity in Metals? Unfortunately, this is a broader topic about the True Essence of Light itself and has no place in this particular book.

7. **Why all dependent Cosmic Objects are floating in one plane around their host Cosmic Object and this plane happened to be around the Equator of the host?**
   If my hypothesis is correct, it would perfectly explain why all Planets orbit around the Equator of the Sun and not in any other, mathematically possible, orbit. After all, Einstein's theories allow for that… Also, if Space is empty, then the rotational direction of the Sun (host) doesn't even matter. In my hypothesis, however, these Planetary Mechanics would be regulated by the faster speed of rotation of the Sun around its Equator. After all, we know that the Surface of the Sun itself is rotating

faster around the Equator. And, if the Space is fully filled with a Substance, then this Substance too must be following the pattern and moving faster around the Sun in the plane of its Equator... Just think of it! We know that any freely floating objects in a stream are being pulled in the areas where this stream is the fastest. In case of rivers, it is normally at the middle of it. So, obviously, the idea of a Substance that would have filled the Space would help to explain these Planetary mechanics too. All the rotations of all Cosmic Objects themselves would be caused by their Electromagnetic Fields. I describe it more precisely in my first book, which is titled – **New Theory of Where We Come from and How the Universe Works**. And yes, I do know that there are moons that are rotating in opposite direction. And no, this will be explained in another book as this also isn't the topic of this book.

8. **How all Electromagnetic Phenomena propagate.**
   Presently we believe that Magnetism and Light do not need a medium to propagate. We believe those things despite knowing that Electricity, which is another Electromagnetic Phenomenon, needs a medium to propagate forward. We were forced to think in this way about Light and Magnetism because we thought that Space is empty void with little to no Particles in it. But, by introducing the idea of the Vacuum as a highly stretched Liquid Substance this problem would be solved, and we could start to work again on the ideas that Magnetism and Light need a medium simply because it cannot be true that one Electromagnetic Phenomenon works differently from another. And Electricity needs a particular type of medium to propagate efficiently. And the same must be with Light. As we know it, Electricity can be transformed into Light, and the Light can be transformed into Electricity. If the Light doesn't need a medium and Particles - Photons

never lose their speed, which goes against any Laws of Physics, then we cannot even explain why it has the existing speed to start with. After all, if nothing can affect the speed of Photons, then there is no reason for them to be instantaneous. This would be as illogical as the existing interpretations of what is Light and Electricity. Famously, Chemists and Physicists completely disagree on how Electrons behave when it comes to Electricity. What is a good explanation for ones, is impossible according to others. And yet, both these mutually excluding theories are in our textbooks side by side… Why!? In my hypothesis, however, the Light and Electricity is an increased level of Energy in Particles which propagates by jumping in a copy/paste manner from Particle to Particle. This explanation has no need for either Electrons nor Photons and doesn't change the chemical structure of any elements ever. But, for that to be explainable, we first need to figure out whether my hypothesis about the True Essence of Vacuum is correct …

9. Magnetism and Gravity.

If Light and Electricity need medium, so does the Magnetism. And if the Space is filled with a medium, then maybe Gravity needs it too. Actually, maybe Gravity is Magnetism which is manifesting itself in a different way here. And this is why we cannot understand what Gravity is to start with. After all, if we have misunderstood the True Nature of Vacuum, we will never understand the Light and Gravity. Frankly, we have no clue what Magnetism itself is. I, however, have a fully developed theories on the True Pattern of Magnetic Field and Magnetism too. Also, I have ready theories about what is and how works Gravity. But the world of science is in deep state of denial right now and is fighting against any ideas that don't support General Relativity or Quantum Mechanics. In order to explain

Magnetism and gravity, I desperately need to change our perspective on what is the Essence of Vacuum. But for that I need those experiments, the ones that I will suggest in this very book, to be carried out by science. So far all science journals I have sent my ideas to have been rejecting them without consideration, let alone carry out any of my suggested experiments.

10. The Dark Matter.

As we know, the idea about the Dark Matter derives from Einstein's hypothesis of what is and how works Gravity. And this, once again, has led to unsolvable problems, where only one of them is known as the Dark Matter. In reality, it is the same story as with Expansion of the Universe and Dark Energy – these strange concepts derive from misinterpretation of some basic, observable facts. These misinterpretations about how things work later led to situations where nothing makes sense, and nothing can be tested and/or understood. And this is why we are where we are with Theoretical Physics today – stuck for some 50 years in the same place. And this book could be our opportunity to break the ice and start moving forward again. Unfortunately for science, most of the existing theories will have to be dropped and we all will have to return to the drawing board. And the True Essence of Vacuum should be the first topic to be fully revisited and retested. The Dark Matter was needed to be introduced because it is impossible to explain the orbital speeds of Stars at the edges of galaxies. However, if my hypothesis about the Essence of Vacuum and Magnetism causing alignments and rotations of Cosmic Objects are correct, then the orbital speed of any Cosmic Object would be determined by the rotational speed of the object itself and the direction and speed of the Space - Substance around it. Since there is such thing as the Centrifugal Effect, the Stars that are larger and rotate faster would have thrown themselves further away from the Center of the Galaxy. Also, since all Cosmic Objects would be surrounded and "tight" by the "Ocean of the Cosmic – Substance" which they also attract, they would be as if tied by it to a certain area like by strings that have fluid properties to them. At the same time, Cosmic Objects would propel themselves in this Cosmic Ocean forward. Without that we cannot explain why they stay in their orbits. Einstein's theory of Space-Time Fabric, no evidence for which has ever been found, doesn't explain the Planetary Mechanics (Mechanics of Cosmic Objects) fully. So, these Stars at the edges would be keeping the whole Galaxy stretched and prevent it from collapsing together because of the Attraction Force which we call Gravity nowadays. And no Cosmic Objects, Galaxies or Clusters of Galaxies, would ever move away from each other or closer to each other permanently. They would only move further or closer for a short period of time which would be followed by the opposite movement in the same way as do Planets around the Sun. So, no Expansion of the Universe and no Dark Matter, I am afraid…

As you can see, a huge range of very important topics will be affected and explained in a different and much easier, logical way, if the Vacuum was understood as it is in my hypothesis. And all those explanations would be within the known and proven Laws of Physics. So, why not to give a shot to those tests that I am suggesting? What is it we can lose? Either it will lead to new advances in Physics, or it will clearly and additionally prove that the existing theories have been true all along. I have to admit that I strongly believe that they are wrong.

More details about some of the topics described above you can find in my first book. The title of it is, as I already have disclosed it, as follows – **New Theory of Where We Come from and How the Universe Works**.

Right now what science needs most is to show the world that any thinking out of the box is welcome. I believe that even if you might disagree with many things expressed in this book so far, you still should agree with the statement that sometimes it is the wrong ideas that trigger a chain of events which later lead to new and amazing discoveries. Of course, any ideas suggested should be logical and have some water in them. Ideally, they should have suggestions for experiments and predictions that would be uniquely possible only with them. And I believe that I have all of that in this very book with respect to the True Essence of Vacuum.

THE TRUE ESSENCE OF VACUUM

CHAPTER FIVE

The full description of what is Vacuum in my hypothesis, suggestions for experiments to test it, and the predictions following these tests.

I would like to start the description of my hypothesis about what is Vacuum with an image depicting it visually and describing the lack of sound in it.

Image 1

Why the sound doesn't propagate in Vacuum and in Space if they are fully filled with a Substance where there would be no gaps between Particles?

Imagine a musical instrument that has strings. If its strings are too loose, they don't vibrate and, for this reason, they don't produce a sound. When they are tightened properly, they vibrate and produce a very good sound. But when you overstretch them, let alone extremely, they once again become nearly silent at some point.

Now imagine that Particles of fluid Substances are like these strings. When the given Substance is extremely stretched, its Particles cannot move and respond to irritation of vibrations of whatever the Cause of Vibrations would be. So, instead of conveying the sound, they actually stop its propagation. Electromagnetic vibrations, however, keep propagating as usual through them because those are completely different vibrations.

The example with overstretched strings of musical instruments explains the mechanics that stops the propagation of Sound in the so-called Vacuum despite it consisting of the same Particles that previously were good conductors of sound. In the case of strongly stretched Particles of a Substance (which serves as a Medium for the Sound) the medium itself gets stretched and this leads to it losing its ability to transfer the Sound. The loss of Sound will happen gradually following the amount of expansion rate of Particles. At the end, the Sound appears to have disappeared completely. Possibly, some very sensitive devices still would be able to pick up the sound even in the Vacuum. As we know, our ears, in the same way as our eyes with respect to the Light, can only perceive a certain range of all spectrum of Sounds around us. So, for this reason the fact that we don't hear sound in a Vacuum doesn't prove that the given area is Matter-free.

Keep in mind that the Medium of Sound itself isn't producing the sound – it is transferring it. But, after the medium has been significantly stretched, it cannot do this job anymore. For this reason, it is perfectly logical that despite sound not propagating in a Vacuum, it still could be consisting of a Substance which Particles are in full contact with each other exactly like the image above depicts it visually. For this reason, all Electromagnetic Phenomena (like Light) still would be propagating perfectly through the Vacuum, as it still would be fully filled with Particles that touch each other at all times and from all sides.

In my theory the Vacuum is "an upside-down version" of the compression of the given Substance. For instance, if you have a sealed container in which you are trying to compress the Substance, then it will lead to Particles collectively push back.

These Particles would be in full contact with each other and with the walls of the container. And the density of the Particles per certain volume will be very high. In case of a Vacuum the Particles of the Substance would be stretched, but they would still be in full contact with each other and with the surrounding walls of the given container. And, contrary to compression, the Particles now would be pulling inwards.

But this I am going to explain later.

THE TRUE ESSENCE OF VACUUM

***

A very important thing to keep in mind when considering my hypothesis – How Particles work in my hypothesis.

I have said this already, but it is important. For this reason there will be no harm if I repeat it once more.

So, in my hypothesis all Particles are like tiny Magnets and they are rotating. Also, they all are holding together at all times. The problem of them moving from place to place is explainable with the claim that, even though they are strongly holding to anything that is the closest to them on all sides, they still can roll over and move from one relationships to others if they are forced to by whatever the force is. This is only possible with fluid Substances. Particles of solid Substances always stay together and roll nowhere unless they are heated up and become fluids in this heated up state. So, in interactions between solid and fluid Substances it would be the Particles of fluid Substances that would roll over the solid ones. Also, exactly the same thing would happen when these solid Substances are being moved through the fluid ones.

And it works in the same way between any two or more liquid Substances. For instance, if we look at Water and Air, they both seem to break apart easily when interacting between them. In a minute I will explain to you why this only appears to be so and what proves that this intuitive perception is wrong. It is our observations of these interactions between fluid Substances which make us assume that there is nothing that keeps their Particles together. But that can be disproved very easily and few of my suggested experiments are designed to prove just that.

What proves that Particles of Substances keep together by some unknown method?

Imagine a large cistern filled with Water. Imagine that this cistern absolutely full of Water only and is perfectly sealed apart from one opening through which you will have to pump the Water out. Now, try to pump the Water out of this cistern from anywhere, be it the top or the bottom of it.

The thing is - if this will be the only opening in the cistern, and no Air will ever enter it, you will fail completely to pump out anything. And this can be possible only if the Particles of Water are holding together to each other and to surrounding walls of the cistern very strongly and at all times. Why is it possible to pump out the Air then, right? That is because the Air is very elastic, and the Water isn't elastic at all. And, by pumping the Air out of such cistern, you simply reduce the density of Particles of the given Substance inside of the given cistern. The idea was shown visually in the previous image. Also, it wouldn't matter whether you are pumping the Air out from the top of the given cistern or from the bottom of it. You still will get out exactly the same amount of Air from it. And that too doesn't fit into the existing theories of what is Vacuum, but it is perfectly logical in my hypothesis.

So, why does the Water break up easily when outside in an Atmosphere? This happens because there is no empty gaps between two liquid Substances. Any two or more liquid Substances work together as one even though one might be far more elastic and the other might be far denser, heavier, and not elastic at all. Whatever appears to be the separation between the Particles of the same substance, simply is replaced and filled with the Particles of another Substance. So, still no empty gaps between Particles, I am afraid. In this way a full contact between Particles of all Substances remains at all times despite it appearing that Particles easily break up if two or more Substances are interacting.

So, if you would fill a cistern with 90% of Water and some 10% of Air, then you will be able to pump out some Water from it. And this will be possible only because the Air will expand for a certain amount and allow for the Water to be pumped out. But it will allow it for a certain amount only. This can only be explained if the Particles of both Substances are in full contact with each other and also with the walls of the cistern.

THE TRUE ESSENCE OF VACUUM

Do you remember those experiments with collapsing railway cisterns? Yes, I am sure that this happens not because of the Atmospheric pressure on an "empty cistern" as contemporary science interprets it, but because of the inside pull of those few Particles that are left inside of the cistern. And this is exactly what I am trying to prove with the suggested experiment that you will find later in this very book.

Please, take a look at the next image.

Image 2

This picture is a screenshot from the YouTube channel "**Tom Brattain**". Title of the video is – "**Railroad tank car vacuum implosion**"

If we tried to pump out the Water from this cistern with no Air in it, it would be close to impossible to get out anything.

If there was 5% of Air inside, some Water from the bottom of this cistern would be possible to pump out.

If there was 10% of Air in this cistern, then we would be able to pump two times more Water out compared to the previous experiment as there is two times more Air inside in this case.

Of course, I have not done any such experiment. I am only hypothesising. After all, I don't have any railroad tanks to ruin...

In my view, the results of these experiments can only be explained if all Particles of all substances are in full and fast contact with each other at all times, and the Particles of Air are more elastic than those of Water. For this reason, the percentage of Air inside the cistern would affect how much Water we can pump out of it.

Borrowed from: https://htgetrid.com/lv/stavit-banki-na-spinu-polza-i-vred-dlya-zdorovya/

Image 2/B

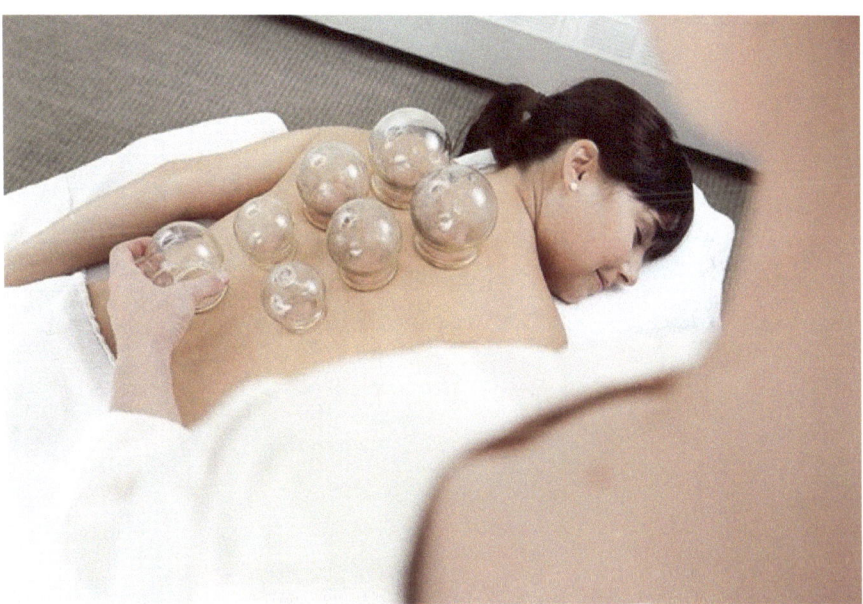

Many people are familiar with these medical methods. The reasons for them and benefits of them are irrelevant for the sake of this book. We are interested only in the True Essence of Vacuum here.

If scientists are correct, the air in these banks isn't sucking the flesh inside of them. Instead, we are told that the tissue of our body is pushing itself inside the banks... But is this true or even possible? This is definitely counterintuitive. The space of the inside of banks clearly seems to be sucking everything inside. So, what is happening, right!? And we cannot accept this idea only because we have assumed that Particles of liquid Substances aren't holding together as do Particles of solid Substances. To fix this, we need to perform some experiments...

The thing is - my hypothesis explains all observations perfectly. Firstly, the heat of an open flame heats up the Air inside the bank. When Particles of Air are heated up, they expand very significantly. Then, when the flame is taken away and the bank has been put on the body of the patient, the Air inside of the bank starts to shrink because it starts to cool down. Keep in mind that Air is very elastic gas and very responsive to impacts of Temperatures.

Anyway, the cooling down process doesn't happen instantaneously, but still it is quite fast. So, since in my hypothesis all Particles are in full contact with each other at all times, they start to pull everything that they have a contact with inside when they are shrinking. This happens because no other Air is allowed inside. The glass doesn't give in. And this is why the flesh of a human being has to and, therefore, it is being sucked inside the banks.

Another small experiment that defies existing interpretations about the Vacuum.

Image 3.

In my view, one thing that proves me right is my experiments with a sink plunger that can be made stuck to a smooth and wet surface. Have you ever asked yourself - Why does it stick to the wall and doesn't fall off? Of course, nowadays we have many different "vacuum holders" that work using the same mechanics, so I could have used a "cleaner object", right!? The thing is - the situation would be identical anyway. And, after all, at school I was shown this phenomenon with a sink plunger. So, why to ruin a good traditions, right!? Wink!

As you understand, above is a picture illustrating what I mean with this experiment! But it tells only a superficial story about my experiment.

So, what was the experiment about?

At school demonstrations the teacher simply makes it stuck to the blackboard. And he asked the class: "Why do you think the plunger is holding there and isn't falling off?"

After some attempted guesses by students, he/she reveales

the official story of the science as this – the plunger stays on the wall because the pressure of the Atmosphere is holding it there by pushing it as there is no Air between the plunger and the wall. It instantly didn't make any sense to me. But who cares, right!? In those days I was happy when the bell, indicating the end of the lesson, rung. I was curious and wondering about the topic in my head for a while but only whn I still was in the class, and then it just vanished from my mind. But it returned to it sometime later and kept returning occasionally. After all, I am a curious person who really wants to understand all things around me fully and so that there are no gaps that make no sense in the existing interpretations. And "the sink plunger story" was one of those that bothered me occasionally ever since I was told about it, as I couldn't agree with the existing interpretaton. But the qusetiion was this – how to prove the existing theories wrong and my interpretation correct! Is that even possible?

Image 4.

Please, take a look at the next image that is meant to explain what is happening according to my interpretation with this sink plunger.

As you can see in the image above, the Particles inside the sealed area of the second plunger are depicted to be stretched, they are holding together, and they are pulling back inwards. This happens because all Particles in my hypothesis are in full contact with each other and with all the walls surrounding them. They are holding with whatever they touch with incredible force. And, for this reason, these inside Particles become stretched and are pulling inwards at the same time when the rubber part of the plunger is pulling itself away from the wall which results in it stretching the inside Particles. This creates two forces pulling in opposite directions. And this pull is stronger than the pull of Gravity. So the plunger get stuck on the wall.

Let us go through all of this once more – So, because the edges of the plunger are wet, it stops Particles of Air from outside to get inside of this sealed area. And because the Particles are in full contact with all surrounding walls that they are touching, and also with each other, they keep pulling the plunger back towards inside. These two pulling forces (rubber of the plunger and the stretched Particles in the compartment between the plunger and the wall as one, united system) are strong enough to keep the whole plunger to the wall despite all of the elements involved being pulled downwards by Gravity.

I explored this "plunger idea" in two situations, which additionally confirmed that I might be correct with my interpretation. I got this plunger stuck to the wall when its inside compartment was fully filled with Air, and then I did the second experiment when its inside compartment was fully filled with Water. And guess what, it behaved very differently when being pulled of the wall. The differences, in short, were caused by the properties of the Substance that was inside of that compartment at the given situation.

Since the experiment was carried out in the same bathroom, at identical altitude, I can safely declare that the pressure on the plunger from the outside Atmosphere had to be the same in both experiments. So, what was different was the way plunger came off the wall when I pulled it. In the experiment with Air the compartment was giving up gradually, as if its inner side were stretching. But when stretching (as I was pulling it off the wall), I had to use more and more energy as the resistance was increasing during this pulling off process.

Clearly, since the outer surface area of the plunger didn't change, and if nothing was holding the plunger back from inside (as the contemporary science tells us), let alone with increasing resistance, then the growing resistance simply doesn't make any sense within the existing interpretation. However, with my own interpretation these observational results (facts) are exactly what we should expect. And only when the inside pull was strong enough it forced the sides of plunger to give up by bending inwards so much that the Air from outside got inside the sealed compartment, then the plunger instantly came off.

When the compartment of the plunger was filled with the Water (only its sealed compartment was full of Water as I let the Water out of the bath after the plunger was stuck to the side of the bath), it appeared to be much harder to pull it off the wall initially. In reality, it only appeared that way at the beginning as it wasn't giving up in the same way as it did when the compartment was filled with Air. This, in my interpretation, is because Water isn't elastic. However, when I reached the same amount of strength, when pulling the plunger off, which was needed to get the plunger off the wall in experiment with Air, the same happened with this experiment too – the sides of the plunger gave in, the Air got inside and the plunger came off the wall.

So, in both cases I needed the same strength at the end to get the plunger of the wall. The only difference was the behaviour of the plunger when it was in this process of pulling it off. In the experiment with Air the area between the plunger and the wall was stretching during the pulling-off process, and it was becoming harder and harder to pull it. The pattern kept steady up to the point when it finally came off. In the experiment with

Water the plunger didn't stretch when it was being pulled stronger and stronger, but it simply came off the wall when a certain pulling strength, seemingly similar to one needed in experiment with Air, was reached. Conclusion? - Water isn't elastic, the Air is. I wonder whether this property of Particles of a given Substance (how elastic they are) contributes to how good conductor of Light the given substance is… It might be. After all, the Light carries also the heat. And the heat expands the Particles of the Substance. But if these Particles aren't elastic, the Temperatures and Light cannot affect the less elastic Substances apart from a tiny amount. For this reason, the Light and Temperatures are traveling much shorter distances in Water and they can only reach certain temperatures there. These Temperatures are much lower compared to those achievable in Air…

In my view, all these experiments already prove that the interpretation of contemporary science about Vacuum might be wrong and might be nothing but a free interpretation of what could be happening here. And these interpretations lack imagination and aspiration to understand it fully.

The existing interpretation is based on our assumptions that Particles of liquid Substances don't hold together as it is with Particles of solid substances. And, as I have said it already, that might be completely wrong. The existing notions about the behaviour and properties of Particles of liquids come from observing them here on Earth. But, we cannot judge about that by observing Water seemingly easily breaking up simply because two liquid Substances, if I am correct, might be acting as one, united system. For this reason, Water in Atmosphere breaks up easily, and the same is with Air in Water. But if we separate these Substances and put them inside a sealed container, and then try to pump them out without letting any other Substance inside of that container, they all of a sudden don't break into separate pieces anymore.

I also said that if a cistern was filled with Water only, we all of a sudden are unable to pump this Water out even from the bottom of it if the cistern is fully sealed apart from the opening at the bottom which we are using for pumping out the water. The same principle can be explored in much simpler way. Take a glass buttle filled with water. Put it to your lips and try to such the Water out of it without letting the Air inside. You will fail at this attempt. The Water will not come out. Why? Is it the Atmospheric pressure that keeps it inside? But you are sucking the Water out and the Atmospheric pressure has no access to this Water... And that can only be explained if my hypothesis is correct – Particles of liquid Substances are holding together at all times as do Particles of liquid Substances. After all, if the Atmospheric Pressure isn't holding inside the Air, it shouldn't be holding inside the Water too. And no pumping strength will get that Water out of the sealed container...

These wrong assumptions (broadly ignored and unexplored by science) could turn out to be a big mistake that has been holding our scientific advances back for many hundreds, if not thousands of years. Clearly, some experiments in respect to this should have been performed to prove it right or wrong beyond any shadow of doubt. But science failed to do this. so, this is up to us then...

Please, look at the next image now.

Image 5.

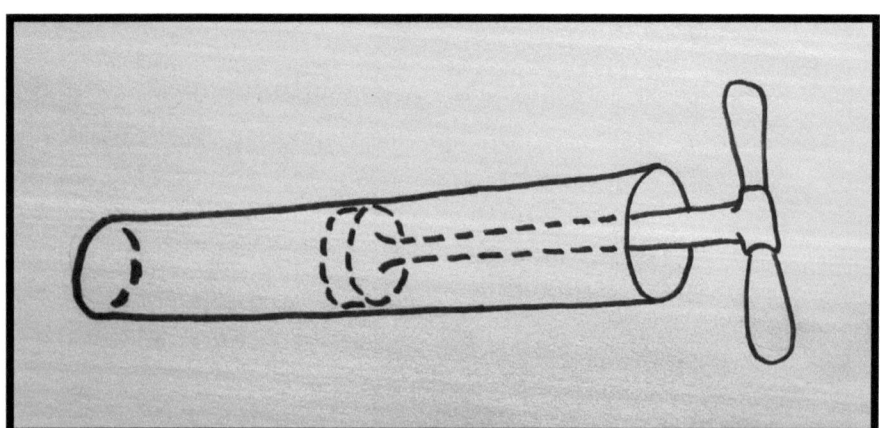

Forgive me the quality of my drawings. Still, it is good enough to make a point…

In front of you there is an imaginary pump. Imagine that its sealed compartment is locked so that no Air can get inside of it. When I was a young boy, I used to play with pump on quite a few occasions when trying to figure out how to fix my bike. I closed the end of the hose and tried to pull the handle of the pump out. It was giving up easily at first, but the resistance kept growing the more I pulled it out.

What does that mean?

If there is nothing pulling the piston back from inside, as contemporary science tells us, then we can safely declare that the only force making the piston to move back inside comes from outside. So, it must be the Atmospheric Pressure. But this Atmospheric Pressure always affects the object proportionately to the surface area of it. And, in this case (especially as it is depicted in the previous drawing) the surface area of the piston of the pump doesn't change. Yet, the resistance keeps growing. Why? If the claim about the Atmospheric Pressure is true, then this growing resistance is completely contradicting the existing Laws of Physics. In my hypothesis, however, the observations perfectly match the theory…

Now, please, look at the next image.

THE TRUE ESSENCE OF VACUUM

Image 6.

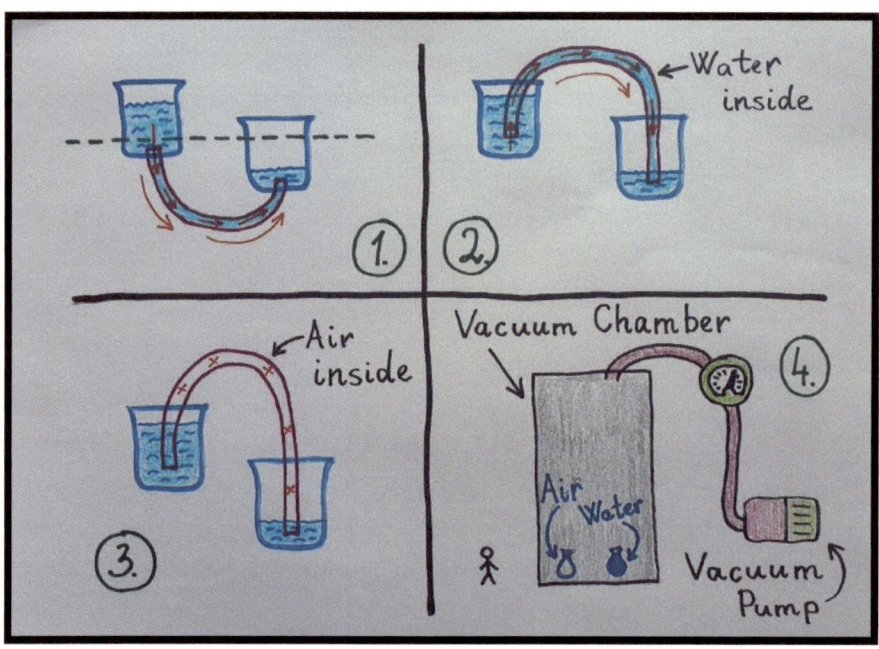

The first part of this image is clear. Water "finds its own level", as "some people" like to describe it! Wink!

The second part of this image is not that straightforward at all. There you see the Water moving from one side to another not because one container is higher, but because one side of the pipe is longer, making the Water inside of it heavier and, for this reason, pulling downwards stronger. And, because all Particles of Water are holding together, and the Air cannot get into the pipe in this situation, the Water with the longer pipe keeps moving downwards stronger and pulls along the Water from the other container that is located higher. Again, this makes a perfect sense within my hypothesis, and doesn't make sense within existing theories.

The third image shows the same situation as the second one, only with one difference – there is an Air in the pipe instead of Water. The Air is much lighter and more elastic. So, even though we can safely declare that the Air in the longer side of the pipe is heavier, it doesn't have the strength to lift the Water. The elasticity of Air also contributes to the lack of flow of Water in this situation.

The fourth part of this drawing shows a Vacuum Chamber with two balloons inside of it. Those balloons must be very elastic and very strong. And, they should only have a tiny amount of Substance inside of them. In one of them there would be Air with the density as it is in Atmosphere, in another Water. And then the Air would be pumped out of the Vacuum Chamber.

As you can see in the drawing, the Air is being pumped out from the top. And, as I said previously, it would not make a difference if the pipe for pumping out the Air is attached to the top or to the bottom of the Vacuum Chamber. In all cases, you will be able to pump out exactly the same amount of Air from this Vacuum Chamber. But that defies, once again, the existing theories of what is Vacuum. After all, if the Particles of Air are not connected, then by pumping them out from the top, as shown in the previous image, we would only be able to suck out a small amount of Air. It should be working the same as it is with pumping out Water in the same way, and if the Air is allowed to get inside the container. You would be able to pump out only that much of Water as much the length of pipe inside of the container can reach. And the same has to be true with Air… But it doesn't work like that at all with Air… You can pump all of Air out regardless of where your pipe is attached to the given Vacuum Chamber and how deep it goes inside of the Vacuum Chamber.

Now, when maximum of the so-called Vacuum in the Vacuum Chamber has been reached, the balloon with the Air should be very expanded, while the balloon with the Water should be as big as it was before the experiment started, provided that there were no Air in this balloon.

If we had measured the amount of Air inside of the first balloon before the experiment, maybe we can calculate how much this volume has expanded. Then, we can build a pump of that kind as shown previously and try to measure how much force we need to expand the same amount of Air by using a pump. At the same time, it will give us a real picture of how large force is needed to expand the Particles of a given Substances at certain stages of expansion.

As a matter of fact, I carried out another experiment with respect to this. but this is a well-known experiment in science. I am sure you are familiar with it already.

In this experiment I had an ordinary glass and a random plastic cover from my fridge. I filled the glass fully with Water, then I put this plastic cover on top of the glass. Then, by gently hold the cover in this position I turned the glass upside down and took off my hand from the cover. And guess what – the water couldn't get out of it. As I said, many of you will be familiar with this experiment already. This is the glass and the cover I had. The glass is filled with Water up to the brim.

Image 7.

Image 8. This is how I put the lid on the top of this glass.

By holding the lid in its place with my finger, I turned the glass upside down and then took my finger off. The Water allowed a tiny amount of Air to slip inside the glass and then it seemed to have sucked the lid back towards itself and from that moment on it was holding the cover fast and neither the Water got out, nor the Air in.

Images 9

In this image the glass is only half-full of Water.

Image 10.

In the next image there is only a tiny amount of Water in the glass.

Image 11.

What is worth mentioning, in my view, is the fact that a tiny amount of Air was allowed to squeeze inside the glass when the glass was fully filled. If there was a tiny amount of Air before turning the glass upside down, then no Air ever got inside of the glass. I thought that it might be because the weight of Water was too high, and it expanded a bit downwards letting some Air inside. When the downward pulling force of the whole volume of Water doesn't reach the level where the Particles of it start to stretch, then nothing moves out or in the given glass. To prove if this is correct I would need a longer tube filled with Water and carry out some extra experiments.

After these experiments I started to make small holes in the cover. And guess what!? Water still was stuck inside the glass. I started with one hole and kept adding them all the time. I kept these holes exactly over the opening of the glass. You had to keep them covered while turning the glass over. Otherwise the Water was escaping through them. When the glass was sideways with respect to the surface of Earth, the Water was pouring out of the glass through these holes. But after it has been turned upside down fully and you removed your fingers from the cover and holes, then no Water was escaping anymore.

Image 12.

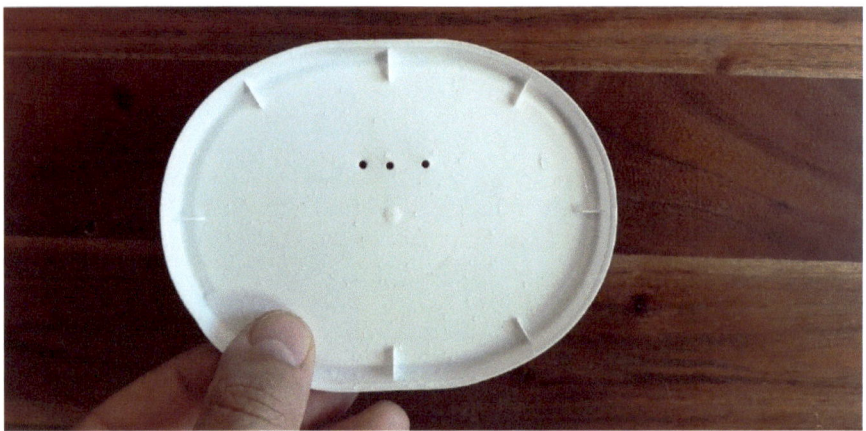

I ended my experiment with quite a few holes in the cover.

Image 13.

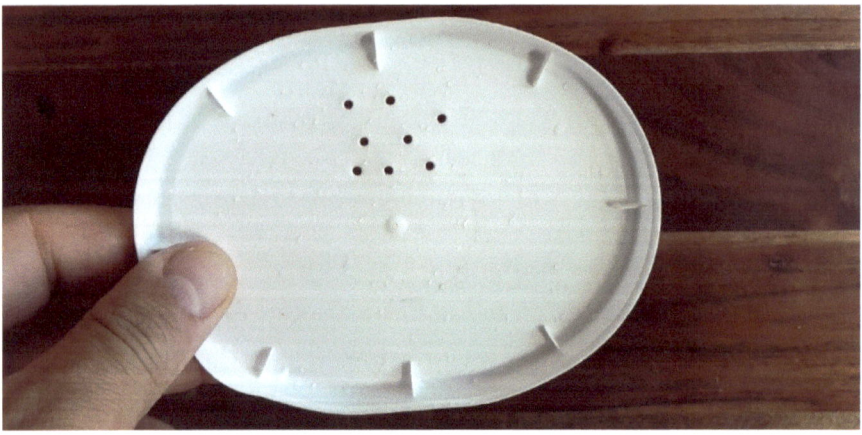

And then I started to notice a difference between no holes, few holes, and many holes in this cover with respect to this experiment. At some point the cover started to stick to the upside-down glass much weaker. It was obviously ready to fall off easily. At this stage it even looked as if there is no longer a full contact between the brim of the glass and the cover, as there was a tiny layer of Water between them instead... Still nothing was moving out of the glass. But one more hole would make it fail.

The thing is, the Air cannot squeeze inside of this glass through the holes if the holes are of a certain size. The size, in order to work (in my interpretation), should be smaller than the natural size of droplets of the Water when it is in a free fall in our Atmosphere. At this size Particles of Water keep together too strongly for the friction between the falling Water and Air to break it apart into smaller sizes.

Holes in the image above were 2 millimetres exactly in diameter.

If the holes are larger, the air would squeeze inside, and the water would start to dash out.

At the same time, when there are many holes even though they are small, then some of them allow the Air to partly squeeze in through them. And for that reason, the Water from inside the glass can move down for this amount. Still, it cannot get out at this stage.

The important thing to keep in mind with respect to this experiment is the fact that the Water will be trapped inside the glass, with holes or not, only when it is fully upside down. When you turn the glass to its side, even if a little bit, the cover instantly falls off or the Water starts to escape through the holes. And that doesn't make sense within the existing explanation of how and why liquid Substances work and why the Water doesn't pour out of the glass in this situation.

All in all, quite interesting experience even if you already know this experiment.

THE TRUE ESSENCE OF VACUUM

***

Suggestions for experiments to prove that there is no such thing as a Vacuum or completely Matter-free areas. Simultaneously, these experiments should prove that Particles of all liquid Substances are in full contact with each other at all times.

One such experiment I already mentioned previously. This is the one with railroad tank and Water in it, where we would try to pump out the Water from the bottom of this sealed container, and it wouldn't move. Nothing else but my hypothesis explains this. Also, for more convincing results, we should try few experiments with this tank when it is filled with a different ratios of Air and Water. But this is not the only experiment with respect to testing this claim of mine.

A suggestion for a second experiment to prove that Space and Vacuum Chambers are all fully filled with a substance, regardless of what this Substance might be, will be explained in a few minutes.

Image 14

In this image I depicted what, in my view, is happening when we are using pumps.

Before we start, a quick reminder about how Particles behave. When they are squeezed beyond the comfortable level for the given conditions, they push back. And when they are being expanded, they pull back because they are in full and firm contact with each other and with all items surrounding them at all times.

So, in pumping process (in one compartment of the pump) we cause the Particles to expand. See that depicted in the previous image. And the pulling force that these Particles create is instantly using the existing opening to pull inside of this chamber Particles from areas outside of the pump which have denser Air. And they will only stop pulling them when the pressure will be balanced between the two sides.

Important! Because of the previously described mechanisms of Substances, it is very difficult to use pumps in Vacuum Chambers and in Space. This happens because the Particles in these places are already extremely stretched. So, for the pump to work in Space and in the Vacuum Chamber, one would need a large and extremely efficient pump to be able to pump anything there from one place inside the other, as this would require a truly great strength and efficiency of the pump. Keep in mind that pumping only works if you can stretch and squeeze the Particles in the given room with your pump. But the Particles in the Vacuum Chamber are so strongly stretched already that it makes the conventional pumps useless down there.

Now we can move on to the actual experiment which would prove that there is no Vacuum. For this, please, see the next image.

Image 15

In this image I have depicted two interpretations for what is happening in Vacuum Chambers. One is what science presently teaches, the other is mine.

<u>The first depiction</u> shows how existing theories explain what is happening in a Vacuum Chamber when the Air, or as much of it as can be possible, has been pumped out. Scientists agree that some Particles are still left in any Vacuum, as they admit that they cannot create a Particle-free place ever. However, they say that those Particles are individually floating around, and nothing keeps them together. So, the Gravity must have pulled most of them to the bottom in the same way as Gravity pulls down Water. Clearly, if this is true, there will be no chance ever to fill any item in this Vacuum Chamber by using pumping. If you were lucky, something could work only at the very bottom of the Vacuum Chamber in this theory. However, we have to admit that it would be only logical to use this story when explaining why there is no Sound in the Vacuum if this interpretation of Reality is true and there are no Particles in Vacuum...

The second depiction shows how I envision what is happening inside the Vacuum Chamber. As you can see, there is no Particle-free areas or gaps between them anywhere. Instead, there are only few Particles left inside this Chamber and they are all extremely stretched. All Particles are in full contact with each other and with the surrounding walls of this chamber at all times. And, as I already explained it previously, the propagation of ordinary sound is impossible in this situation because the Particles are too stretched. Also, and that was explained by me too already, it would be extremely difficult to pump anything inside of this so-called Vacuum. **But not impossible**.

As I said, for pumping to work in a Vacuum Chamber and, indeed, in the Space itself, we would need the most efficient pump we actually can build. And the compartments of this pump should be pretty large too. After all, if these Particles were stretched by the strongest machines that we can built for the job, it would be arrogant for us to even think that it would be easy for us to overpower the incredible stretchiness of Particles achieved by the machines. It would be as silly as when we allowed some large truck to stretch a very strong piece of rubber and keep it in this stretched position, while attempting to stretch it further by the power of our own muscles...

A very simple way to prove my hypothesis in Vacuum would be by taking a conventional pump inside of it and pump all the Air out. Then, we would try out whether the piston of this pump works in there without actually pumping anything. Then we would close the end of the pipe of the pump and try to pull out the piston. If the piston resists, then there must be something holding it back. My hypothesis explains exactly what. The existing theories cannot explain this if that is truly happening.

Remember, if we want the pump to work in the Vacuum Chamber, we have to stretch the already highly stretched Particles in this Vacuum Chamber by even further expanding them inside the main compartment of the pump. Clearly, we will need a very effective, machine-powered pump to achieve that. This explains why conventional pumps fail to work in Vacuum. My guess is that conventional pumps become very stiff to operate and hard to pull in a Vacuum and they also fail to do the job. And this explains how the failure of these pumps, in the same way as the lack of the Sound in Vacuum, doesn't prove that Vacuum is Matter-free area.

However, there is another, more simple way to prove that Vacuum Chamber is filled with Particles from one side to another without any gaps between the Particles, exactly as shown in my previous depiction.

Instead of pumping anything, we would take inside the Vacuum Chamber some specially prepared, soft sided containers. They could be basketballs that have an opening somewhere on the side, which can be very tightly sealed from outside. In fact, it could be any soft container of any size and form, as long as they can be very tightly and effectively sealed while inside the Vacuum Chamber.

So, what we have to do to test my hypothesis is this - We should take these items inside of the Vacuum Chamber with their openings open. Then we should pump all the Air out of the Chamber. This will ensure that inside of these balls there also will be the same density of Particles (or lack of them) as will be in the Vacuum Chamber in general. Ideally, a man in a Spacesuit (or a robot) would stay inside of the chamber. Only that robots take time and money to build them. So, this man or a robot would take these basketballs, ensure that they have a form that makes them look as if they are fully inflated, and then he would very tightly close the openings of these balls with some form or sealant or something. Whatever does the job is good.

This process and results should be similar to sealing a cut basketball in conventional Atmosphere. If the basketball is cut, and if you hold it accurately so that it stays in a form that it has when inflated, and then you very securely seal the cut from outside so that there is no leaks anywhere and the sealant holds the cut very efficiently closed, then this ball will feel as if it is partly inflated. In essence, if you stepped on such basketball that was sealed in a way just described, you would feel the Air in it. And this is exactly what I would like to do in a Vacuum Chamber or in the Space with such balls or other inflatable items.

Now - **the most important bit of this experiment!** If there are no Particles in this Chamber, as it is the claim of the existing theories, then sealing these basketballs inside of these chambers by using the previously described method would mean absolutely nothing. In other words, they still would remain empty. This means, whoever would step on them would fully flatten them as there is nothing inside of them in the same way as there is nothing outside of them, right!?

However, if the Vacuum Chamber is fully filled with extremely stretched Particles that keep together with each other and with the walls of the chamber itself (and all other objects that they touch) at all times (as I am suggesting in my hypothesis), then sealing such objects would make them behave like partly inflated objects that were sealed in the same way as in conventional Atmosphere. This means, they would appear to be nearly full when stepped on.

The important difference between two basketballs filled in this way in different environments would be the elasticity of the ball when stepped on it. When a pressure would be applied to the ball sealed in Atmosphere, it would feel very elastic. When a pressure would be applied to the ball sealed in the Vacuum Chamber, it would feel very stiff or even hard and not giving in. this would be because all Particles in this chamber would be very highly stretched and, for this reason, they would have very inelastic properties to them. This would lead to lesser bounciness of the ball in the chamber as well.

When the Air would be allowed back inside the Vacuum Chamber, this ball would completely collapse showing that they are close to being absolutely empty.

And this experiment and behaviour of these balls would prove that there is no such thing as a Vacuum, as all these predictions would only be possible if my hypothesis is correct. And who doesn't want to know the True Essence of the Vacuum, right!?

And if my hypothesis is correct, we can safely return to the idea that Light, Magnetism and Gravity, whatever they might be, all might need a medium to propagate after all, as now we would have put back on the menu the Substance through which these things might be propagating. The possible positive result of this experiment would firmly confirm that Space and Vacuum are fully filled with Particles that are in full contact with each other at all times and that there is no such thing as Matter-free Vacuum anywhere ever.

Of course, one should definitely carry out this experiment also in Space. And, if these balls do the same thing up there too, then we can bring them back from Space and measure how much those Particles are stretched up there and what the Cosmic Substance consists of.

There is another way to achieve the same results as with sealing basketball. Instead of basketball we could take inside specially prepared balloons. Inside of these balloons there should be some light structures that would keep them stretched and with large inside room. Also, there should be a mechanism to seal the opening of this balloon by a remote control. So, when such balloon would be inside the Vacuum Chamber, we would remotely ensure that its opening is securely closed. Then we would ensure that something heavy is put on this balloon to observe its behaviour. As the inside structure of it would be very weak and fragile, it would easily give up or brake or collapse like a house of cards, depending on how exactly we made it. And, once again, the balloon should feel and look partly inflated. But when the Air would be let back inside of the Vacuum Chamber, this balloon would collapse showing that it is very close to being empty.

After all of these experiments the right question to ask would be this – why these objects feel inflated, if there is no Substances in Vacuum?

And these are not the last experiments to prove that Vacuum, as Particle-free area, doesn't exist.

However, right now it is a good moment to look at one of those observable facts that led to contemporary assumption about the Vacuum as a Matter - free area. I am talking here about all objects falling in Vacuum at the same speed versus them falling at different rates in our Atmosphere. This was the fifth point on that list that I provided at the beginning of this book when describing all facts that led to assumption that the Space and Vacuum Chambers are Particle – free areas.

Why do all objects fall at identical rate in Vacuum!?

The thing is - this fact of all objects falling at the same rate in a Vacuum appears to be a very strong argument to support the idea that Vacuum indeed is a Particle - free area. This is perfectly logical after all. Unfortunately, this phenomenon can be explained in my hypothesis too.

Image 16.

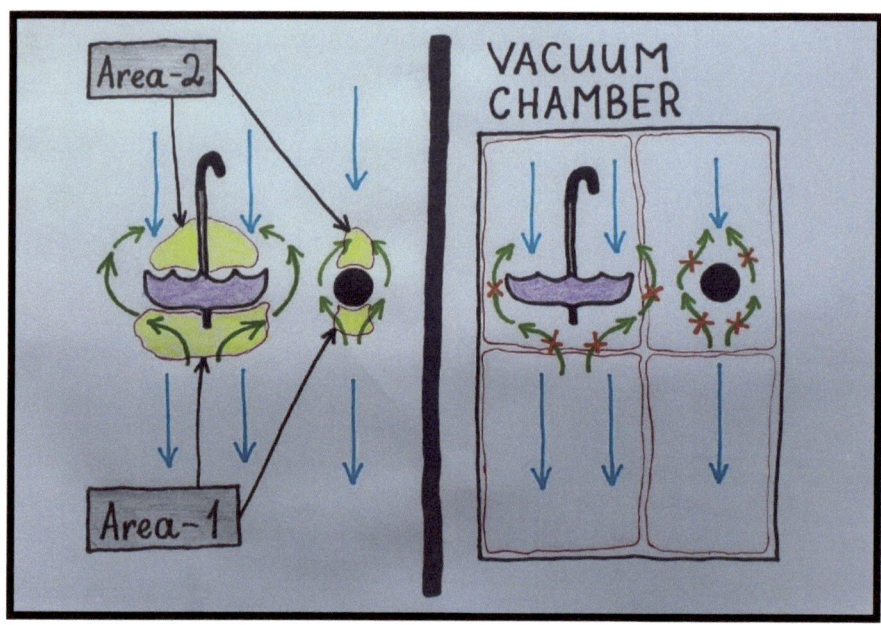

With these images I would like to explain what causes the different falling speeds of objects in conventional density of our Atmosphere versus the Vacuum as it is described in my hypothesis. And the same mechanics would apply to Space as well.

The depiction on the left shows a conventional environment. **Area – 1** in it represents (very approximately) the area where the given falling object has slightly compressed the Particles of the surrounding Substance. The fact that they are compressed compels them to push back in all directions. It means, this area of Atmosphere also is slightly pushing upwards the given falling object. This slows down the falling object. And, since the total surface area of the umbrella is much larger than the area of an iron ball, which is depicted next to the umbrella, then it should be clear that umbrella is being slowed down stronger than the ball is.

Area – 2 in the previous depiction represents the area where Particles become more expanded because of the surface area of the falling object. For this reason, this area is slightly pulling everything towards it, including the umbrella or any other falling object. Clearly, that too slows down the rate of the fall. And, once again, the ratio between the surface area versus the weight of the object matters. Obviously, depending on the surface area versus the weight of the object in conventional density of our Atmosphere determines at what speed the object will be falling in the given conditions.

Clearly, all this is perfectly fine in the environment of our Atmosphere as Particles of it here have a density that allows them to be expanded or compressed easily. And this is precisely what is missing in the Vacuum. So, despite the Vacuum being fully filled with the same Substance, its Particles would be so stretched there that there will be no room for them to be stretched further or compressed. And, for this matter, the mechanics that slowed down both of these objects in conventional density, would have no effect whatsoever in the Vacuum.

In **the depiction on the right** you can see the Vacuum Chamber. Once again I tried to depict the Particles there in a highly stretched state. Clearly, if that is true, then any falling object cannot cause a compression of Particles under it or the expansion of them above of it. This means, nothing would be slowing down the falling object. This leads to the phenomenon that all objects fall at the same rate in a Vacuum despite Vacuum being fully filled with a Substance.

A Substance of lower density has lover viscosity (higher fluidity) too. If we put all these aspects together, and the density of the given substance in a Vacuum is as low as possible, then we can perfectly understand and explain this phenomenon with my hypothesis.

When I was thinking about this, I realised that something similar could be true with our Atmosphere if it were compressed to a maximally possible degree. After all, you wouldn't be able to squeeze or stretch Particles in this environment at all. So, the factors that affect the speed of falling objects in conventional density of Atmosphere would no longer exist in the highest possible density too. This means, that objects could be falling at the same rate at the lowest density possible, and also at the highest density possible. Of course, there would be a huge difference between the rates of fall in these conditions. And objects that have lesser density than maximally compressed Air, would not fall at all...

Clearly, this is a hypothesis right now. Unfortunately, a very compressed oxygen turns into very cold liquid and is quite challenging to make, especially knowing that we would need a large container with windows for observations to carry out any such experiment to test this claim.

Suggestion for a third experiment to prove that what we call a Vacuum is nothing, but an area fully filled with a Substance even though its Particles are very stretched.

We are told that those railroad tanks are crashed by the Atmospheric pressure rather than from the force pulling inwards, as there is nothing inside of them according to Science. And scientists have even calculated how strong this Atmospheric pressure is. According to them, it is 10.3 metrical tons per each square meter. And this is where an idea about testing their mathematics and interpretations emerged.

Image 17

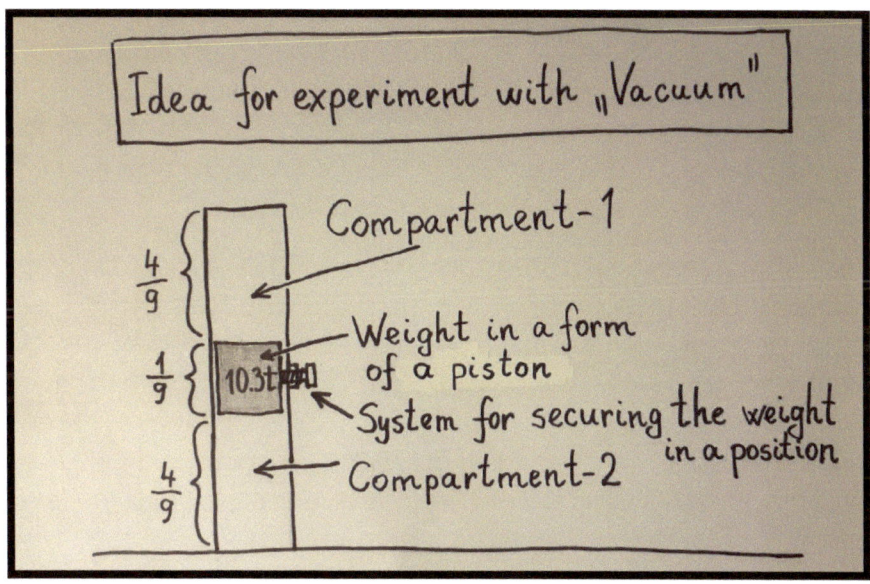

In this image I have depicted a device that is like a large cylinder that can be fully sealed. The area of its circle inside this tube would be precisely 1 square meter. Inside of it there would be a weight designed as a piston that would be 10.3 tons heavy, as this is the weight of the Atmosphere on each square meter...

Since the Atmospheric Pressure is said to be 10.3 tons per 1 square meter and Vacuum doesn't have any pulling power whatsoever, then it should be perfectly reasonable to assume that if we pump all the Air out of the upper compartment of this cylinder, then this weight should stay right in the very position where it was put at the beginning of the experiment. Namely, it should be somewhere at the middle of that cylinder and fixed so that is doesn't move anywhere. After that, the existing Atmospheric pressure has to be allowed in both, upper and lower compartments. Then the system for securing the weight in a certain position would be removed so that the piston can move freely up or down. This would result in the weight moving down for a certain amount. But, because this would cause the Particles at the lower chamber to get squeezed, and Particles of the upper chamber to get stretched, the weight would stop at certain position without reaching the ground. Of course, this will be true only if the cylinder of the experiment is long enough and the piston/weight is designated precisely enough so that Air from any of chambers cannot squeeze into the other one.

So, the Atmospheric pressure in both chambers would be exactly as it is all around us in normal conditions. Now, if we pumped out all the Air from the upper chamber, the Air pressure of the lower one should push up the weight exactly to the level where it was right at the beginning of the experiment. This means, no matter how much Air we pumped out from Compartment-1, the weight would never raise above the level where it was put before the experiment started.

Once again - To ensure that everything is correct, the weight would be fixed in an agreed position at the start of this experiment. Both compartments would be allowed to naturally fill with the Air at the existing Atmospheric Pressure. Then both compartments would be fully sealed. When that is sorted, the Air from the upper compartment would be pumped out.

My prediction of this experiment is that by pumping the Air out of the upper chamber, the weight would move much higher than it should according to the existing theories. So, if I am correct, the so-called Vacuum will be able to pull the given weight upwards way beyond its starting position. Possibly, up to the very top of it or close to it. This would prove that either I am correct, or the Atmospheric Pressure has been calculated wrongly.

I also have another prediction with respect to this experiment that would be impossible within the existing theories about the Vacuum. And my prediction is this – if we gradually and evenly pumped out all the Air from both compartments simultaneously, and both compartments were of identical size, the weight would get positioned right at the middle of this cylinder when full Vacuum were achieved in both chambers. In existing theories it should fall to the very bottom as nothing would be holding it back…

It would be interesting to carry out more experiments of similar nature, just to check how things work in different configurations. The next image will give you a glimpse into one such idea.

Image 18.

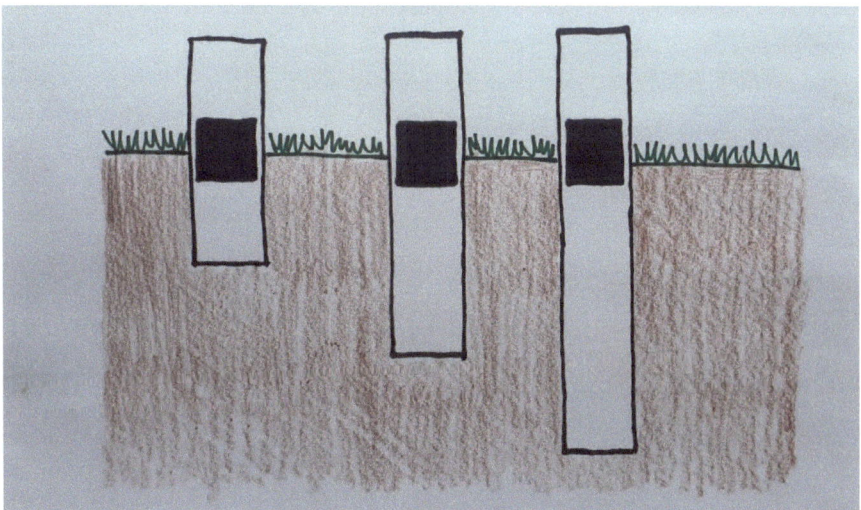

This image simply shows the same experiment as previously, only in a simplified way. As you can see, the size of the bottom compartment is different in each case. If the Particles in the lower compartment aren't that stretched, they can allow the weight to be lifted higher when the Air is being pumped out of the upper compartment. At the same time, the Atmospheric Pressure of the lower compartment still should remain the same if the weight remains at the same altitude. Also, the upper compartment might be enlarged to differentiate the results and broaden the variations.

To make this experiment easier and more feasible, the scale of this experiment might be much smaller. All we have to do is to calculate how big the Atmospheric pressure is on 1 square centimetre, for instance, and then build a device to measure all the effects.

Conclusion!

I believe that this question about the True Essence of Vacuum, as I have explained it at the beginning of this book, is of utter importance in science. It would literally turn over large number of existing theories and open doors for new ones and liberate us from unsolvable problems, such as Dark Energy, Black Holes, what is the True Essence of Light, and allow to have better explanations of many others. Clearly, this is a great topic that needed to be addressed long time ago. Right now it is in your hands. Right now it is you who could help the science to move off the dead point and cause a new revolution of ideas and lead to new breakthroughs in Theoretical Physics. Or not! It is up to you! So, tell me, which side of the history you are right now!?

***

I already wanted to finish this book, when I realised that I can and should address one more question that is very directly connected to the idea of Vacuum. And that is about the Time Dilation or the huge blunders that it contains.

How is Time Dilation related to the topic of Vacuum? Einstein's idea about Time Dilation is very directly connected to another of his idea, and that is the Space-Time Fabric. According to Einstein, the faster one moves through Space, the slower moves his Personal Time. And that originates from one of his so-called thought experiments that led to ideas about the Space-Time Fabric itself. And it is this experiment and conclusions made from it is the question I would like to address right now.

In my theory Space-Time Fabric doesn't exist. I believe that there is a good reason why it hasn't been ever experimentally proven as you cannot find a physical evidence for something that doesn't exist. All the so-called arguments for it are indirect facts that can be explained with different mechanics of Reality. This Space-Time Fabric, and Gravity bending it, have created more problems than they have solved.

Since I have extensively addressed this question in my first book (**New Theory of Where We Come from and How the Universe Works**), I will only look at the blunders of the Time Dilation here.

THE TRUE ESSENCE OF VACUUM

CHAPTER SIX

The Blunders of Time Dilation.

This nonsense, as I said earlier, started with one of the so-called Einstein's "thought experiments". Einstein imagined himself riding on the beam of Light at a speed that is close to the Speed of Light. And he was riding away from a Tower Clock. He realised, that he would see that clock ticking slower. And this is correct, by the way, as Information from objects, in the same way as Light, needs time to reach us. But since we would be moving away from the source of this Information at the speed which affects the time needed for it to reach us, we would experience this information reaching us with increased delay. By the way - Information moves at the Speed of Light too.

So, Einstein realised that moving away from this clock at this speed would cause him to see that particular clock on that tower as if ticking slower. And the faster he would be moving away from it, the slower it would appear to be ticking. So, he decided that this is enough to declare that because of what he is experiencing, the Time itself, actually, must be moving slower for him in Reality compared to those who are left standing at that Tower Clock...

When I am listening to this I always ask myself - why this world is so unjust. When believers in Flat Earth theory say that the horizon of the sea appears flat to them, and therefore the Earth is flat, everybody laughs. When Einstein does virtually the same thing with respect to this "appears slower, so it is slower" thing, he receives a round of applause... Why? Why the double standards?

So, following this story Einstein declared that Time moves slower for those who travel faster through the Space... A Time Dilation for you...

> ***Funny fact*** *– with respect to this claim, up to this day, all scientists admit this – we have no clue why there is this correlation between the speed of movement of objects through Space and their Personal Time compared to others who are stationary. Luckily, I have an answer to this question. And that is a very simple answer – This is because* **Time Dilation doesn't exist!**

The main problem with this Time Dilation is, however, in realisation that Einstein, when travelling in his thought experiment on the beam of Light, could have imagined himself holding a clock in his own hands too. And that clock of his would appear to those standing still at the Tower Clock as if it is ticking equally slower and slower... For this reason, following the same logic, Einstein and all scientists had to declare that Time itself for those standing at the Tower Clock is actually also moving slower (Relativity for you...), while Einstein's Personal Time should be much faster than the one of those stationary observers from their point of view. And in this way all effects would have cancelled out and no differences would be experienced by anybody whatsoever. And at this point I would like to say it very loudly and clearly – this would be perfectly correct within the idea of how the Relativity works. Unfortunately, on this single occasion, all scientists of the world have abandoned the ideas of Relativity and don't want to see them applied here. Why!? I don't have an answer to that!

The story grows even more bizarre if we look at the possibility where Einstein would be moving towards the given Tower Clock and also holding a clock in his own hands. In this situation, the time on the Tower Clock would appear to be moving faster to him than his own personal clock does... Again, following this insane pattern of "whatever appears to be true in certain conditions is declared to be the permanent reality...", he should declare that the direction of movement with respect to observers affect how his Personal Time would be moving with respect to the given observers.

So, Einstein had to declare that his Personal Time, when he is moving away from somebody, is moving slower compared to those he is moving away from, while if he is moving towards somebody then his Personal Time is moving faster compared to those towards whom he is moving to. Yes, if this idea was explored from all possible situations honestly, anyone had to come to these absurd conclusions. Instead we are stuck with lies which are based on only one situation which is interpreted very partially and applied to absolutely all situations. To call this claim and conclusions about the Time Dilation unscientific would be a huge underestimation.

You simply don't know about these failures of this idea because no scientist ever explores it independently and from all sides, let alone challenges this insane claim of Time Dilation. And trust me, this is far from being a whole absurdity that this idea entails…

Unfortunately, scientists decided that this insane claim can only be true in one direction, and absolutely ignored this abominable problem of this theory from all other sides, as if those other situations don't even exist. Scientists like to repeat to us this: **Science doesn't work like that! Science sticks to the scientific method…** In short, this method demands that we look at all possible situations and, **after eliminating whatever is impossible, we accept the only option left regardless of how improbable it might appear.** Unfortunately, in case with Time Dilation, nothing of that was applied. Einstein and all scientists are only looking at one situation only where Einstein is moving away from the Tower Clock and from his point of view only. In reality, there are at least three other situations that anyone has been ignoring for more than 100 years for no explainable reason whatsoever.

Those other situations are these: 1) when Einstein is traveling away from the Tower Clock, observers see the clock that he might be holding as ticking slower, so, the Personal Time of observers must be moving slower… 2) Einstein is travelling on the beam of light towards the Tower Clock and sees it ticking faster, therefore, his Personal Time must be moving faster… And finally, 3) observers at the Tower Clock watch approaching Einstein on a light beam with a clock in his hands and see it ticking faster, therefore their Personal Time must be moving faster...

So, what are you waiting for!? Go and challenge your science teacher with this! They have no answers to this disgraceful fact!

Also, it would be very much in the Spirit of Relativity to consider both (all) sides and in all situations and how the travelling Einstein and the observers would see each other's clocks and only then they could come to any conclusions. But that would ruin the claim about the Time Dilation, and this is such a nice theory (belief of scientific religion) to have, right!? No scientist is going to give it up willingly as it appears...

There is no actual proof for the Time Dilation apart from the claim which all scientists chant – it took the genius of Einstein to realise that this is true... In reality, in days of Einstein there wasn't even any experiments that would prove it. But it was already accepted...

And scientists do it all the time. They only use what suits the already accepted theories and completely ignore what proves those theories wrong. The same is true with the Essence and Speed of Light, Big Bang theory, Gravitational Lensing, Quantum Mechanics, Doppler's Shift and Redshift, and many other ideas in Theoretical Physics...

So, no Time DILATION I am afraid...

Whenever I challenge scientists on this Time Dilation thing, I am often being asked one contra question – **Do you really think that all those highly trained scientists all around the globe and for more than 100 years were all stupid and wouldn't have realised that something is wrong there long time ago, hah!?** And this is often accompanied by other questions – **Who are you? What is your qualification?** I have never stopped wondering how that can be that all these extremely intelligent people genuinely believe that with these contra questions of theirs they have given me a good argument that proved my arguments wrong. Sometimes it becomes especially funny.

When I sent my arguments to **Quora.com**'s **Physics Space**, they rejected to publish it by branding it – a pseudo-science... How ironical, I thought! I have proved with very solid and logical arguments that Time Dilation is nothing but a pseudo-science, and my arguments were rejected by being labelled as pseudo-science... This is like in that old Russian joke and a saying in one form which goes like this – **The one shouting "catch the thief" louder of all is the thief itself...**

This saying contains a great wisdom and points out to some typical patterns of human behaviour. This method people very often use to defend themselves by blaming others about what they themselves have done. And this is precisely what **Quora.com** did with my idea - they branded it pseudo-science to avoid challenging the blunders of already accepted theories.

After scientist and people close to science have asked me these questions, it is only me who realises that any arguments will be ignored without listening to them and I will never be given a chance to explain them. Instead nearly every single scientist will say that Time Dilation is true because no scientist has challenged it so far and that I am not formally educated to raise any questions to start with. And nobody will care about those facts that fully prove that Time Dilation as absolutely absurd idea. Once again, I have to remind to you that this is precisely **Emperor's New Clothes tale in action** and has nothing to do with science. If scientists cannot debunk an argument and instead turn to finding faults in you, then we don't need such science to start with. And that is happening all over the place in contemporary Theoretical Physics.

Some people happily declare that Time Dilation is a proven fact as GPS works correctly only after Einstein's Time DILATION has been included into calculations. In reality, atoms in higher density (closer to Earth) are more squeezed and, for this reason, they operate slower. To understand this correctly, imagine a person moving through a Water compared to the same person moving through Atmosphere at the same altitude. With respect to the speed of atomic clocks the same mechanics apply, and higher density slows them down. And, for this reason, even the Light (and all other Electromagnetic Phenomena) in higher density propagate slower too. So, no need for Time Dilation to explain this... Even Stars are affected by the density. Yes, they too are affected by the density of other Stars around them. This is why Stars at the edges of Galaxies move to fast according to our current theories and we cannot understand that. Dark Matter for you...

Now, let us look at this GPS story closer. To understand how the so-called Time Dilation is involved there, let us first look at what knowledgeable scientist says about it.

Below is what **Dr Ludwig Combrinck** (University of Pretoria | UP · Department of Geography, Geoinformatics and Meteorology) said on **www.researchgate.net** with respect to the speed of clocks of Satellites.

The actual question was this:

> **Are moving clocks really running at different rhythms just because of velocity?**
>
> *Well of course they do, there are 2 effects, one due to gravity and one due to speed. Considering for instance the GPS satellites, the clocks onboard the GPS satellites experiences relativistic shifts, which have both a constant and time varying components. This constant component is compensated for by introducing a fixed offset, which lowers the frequency of the on-board oscillator.*
>
> *The fact that there is some orbital eccentricity and the higher order terms (quadrupole) of the Earth's gravity field are primarily responsible for the components that vary with time.*
>
> *For instance, if two clocks are located on an equipotential surface, the rates of the clocks will be the same. If however one clock is moved to a height of 1 km, their rates would differ by about 1×10^{-13}. Earth's quadrupole's effect on the potential at the GPS satellite*

is approximately one part in 10^14, so in the case of the potential at the height of the GPS satellite, the contribution of the quadrupole can be ignored in most cases (GPS orbits are high enough to be nearly Keplerian); there is also no centrifugal component so that the gravitational potential at the GPS satellite can be easily calculated. **The total gravitational frequency shift of the clock onboard the GPS satellite can then be calculated as 45.688 microseconds per day**. If you ignored this, you would have a one-way navigational error of 13.697 km per day.

Secondly when we consider the speed of a GPS satellite, (~3874 m/s), Einstein's special theory of relativity needs to be applied. **The time dilation effect causes the GPS satellite to appear to run slow by about 7 microseconds per day**.

The GPS satellite clocks are adjusted for three constant rate corrections before launching them into orbit, **the satellite clock has a higher frequency in orbit than on the ground** (read geoid, where the clock frequency should be 10.23 MHz) and its proper frequency is therefore reduced to 10.229 999 995 43 Mhz.

So the nett effect is that between the faster clock in the weaker gravity field due to orbital height, and the slower clock due to the clock's in orbit speed, the gravity effect wins.

One can calculate this very accurately, and it works perfectly, otherwise we would not have sub-cm accuracy in positioning using geodetic quality GPS receivers.

If you paid very careful attention here, you should have spotted that there are two effects that affect the Actual Time of Satellite's clocks. One is Gravity (density), the other is the speed of this Satellite in orbit. And the effect of the Gravity is significantly larger according to this scientist. And I can confirm that the effect of clocks moving faster in high altitudes (low densities) has been very reliably proven experimentally. This is why scientists say that our heads are aging tiny bit faster than our legs...

The thing is, however, that there are many aspects that we still don't understand about Gravity. What if the effects of Gravity on atomic clocks change following different patterns to the ones we are using right now in our calculations? In that case, the expected rate of the speed of atomic clocks at the altitudes of Satellites, along with the results obtained, would be calculated wrongly to start with. But if that is the case, then we might not need to include the so-called Einstein's Time Dilation into these calculations, as with a different pattern for calculations there would be no discrepancy whatsoever and we would be happy without Einstein's ideas. Unfortunately, we cannot calculate each of these effects separately. We cannot even calculate the speed of atomic clock of stationary Satellites as they would instantly start to fall towards the Earth. So, we simply have no clue how the given altitudes affect the clocks. We assume we do. Scientists think that the effect is linear, meaning, the further the clock is from Earth, the faster it moves. But this might be wrong as we don't have any experiments confirming that. As I said, such experiments are impossible.

Also, this is not true that scientists needed Einsteins General Relativity to fix this problem with the wrong time on Satellites. It is more likely that they simply figured out that there is a discrepancy in their existing assumptions and calculations based on them, and they calculated how big this discrepancy is. Then they adjusted all systems so that it doesn't have an effect on our GPS devices here on Earth. And it was only an unfortunate chance that this discrepancy turned out to be similar to some of Einstein's predictions about how the speed of movement through Space would slow down the Personal Time of the traveller. And at this point scientists declared that they needed to apply Einstein's theories to fix the discrepancy and declared that Einstein was right again. In reality, nobody even knows whether the existing pattern for calculating the effects of Gravity on atomic clocks is correct to start with.

There is another experiment that is said to have proved the Time Dilation. It was carried out on the plane. An atomic clock was carried around for some time in a plane, and scientists declared that they are able to confirm that Time Dilation is true. Unfortunately, I don't know precise information about what exactly has been measured and whether the same approach as with Satellites has been used or not. Also, there should be more experiments done with respect to this, but those clocks should be placed not only at the front of the plane, but also at the back of it, to check whether that also affects the readings or not.

Of course, there are some more experiments about it, but they all involve observations of subatomic Particles in certain conditions. As I said, their behaviour in different conditions don't prove anything. Those results might be perfectly explainable without the Time Dilation.

All in all I can say that I would have nothing against the Time Dilation as such, if it weren't based on absolutely partial and absurd claims, as explained previously. Neither Time nor Space are physically existing things that can be stretched or affected in any way imaginable. Atomic clocks do not show a Universal Time. They show the rate at which atoms are pulsating (oscillating???) in the given conditions of density. And that has little to nothing in common with the concept of Time.

More about Space and Time (and many, many other topics, including superficial look at the Vacuum) you can find in my first book – **The New Theory of Where We Come from and How the Universe Works.**

In this book, among other things, I also explore the possible mechanics of the Cosmic Substance and how exactly it affects the Universe and even our weather. Clearly, if the Space is filled with a Substance, it must be made of something else than those gasses that form our Atmosphere. That Substance should be even more elastic and lighter than these gasses. All in all, I believe that that book is also quite an interesting read.

Once again, the True Essence of Vacuum is absolutely imperative for explaining Magnetism and Gravity. And if you have bought and read this book, I would like to share with you the images from my work on Magnetism already now. There will be no descriptions about them at this stage, but I hope you will like them. In case you are interested, a superficial description of where these explanations and depictions of Reality about the True Pattern of Magnetic Field come from can be found in that book of mine which I have been quoting in this book a lot…

Enjoy!

CHAPTER SEVEN

The images depicting the True Pattern of Magnetic Field

THE TRUE ESSENCE OF VACUUM

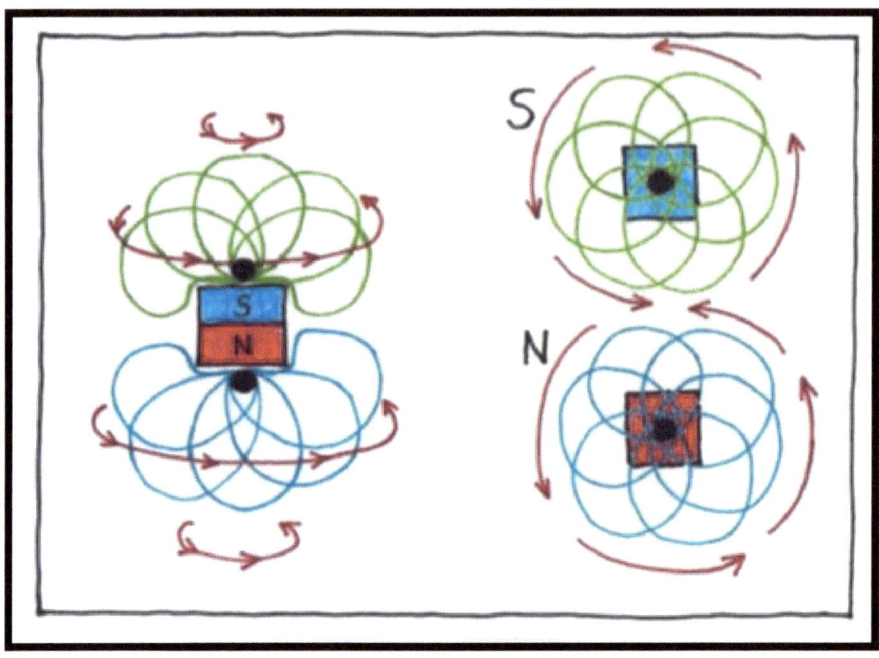

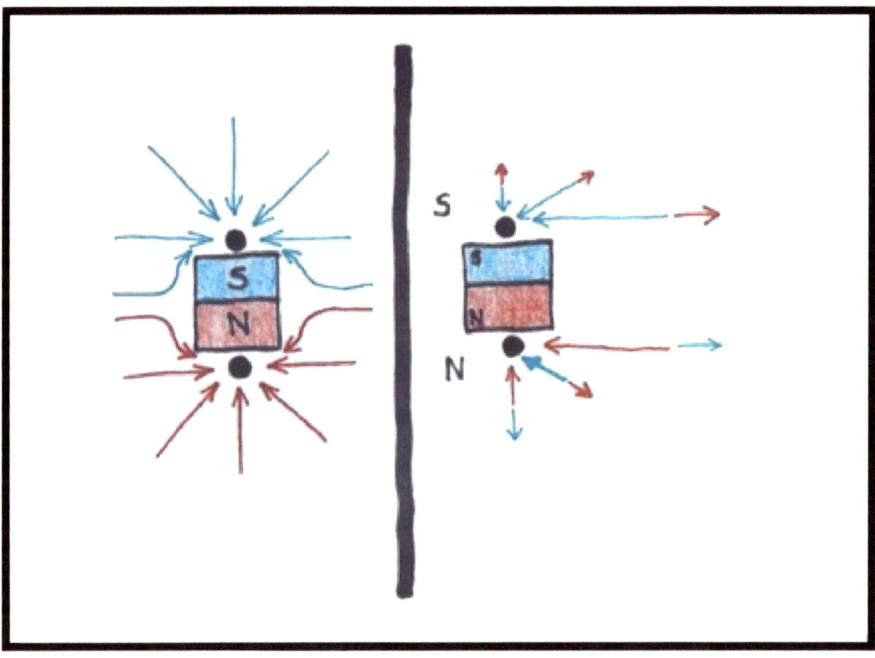

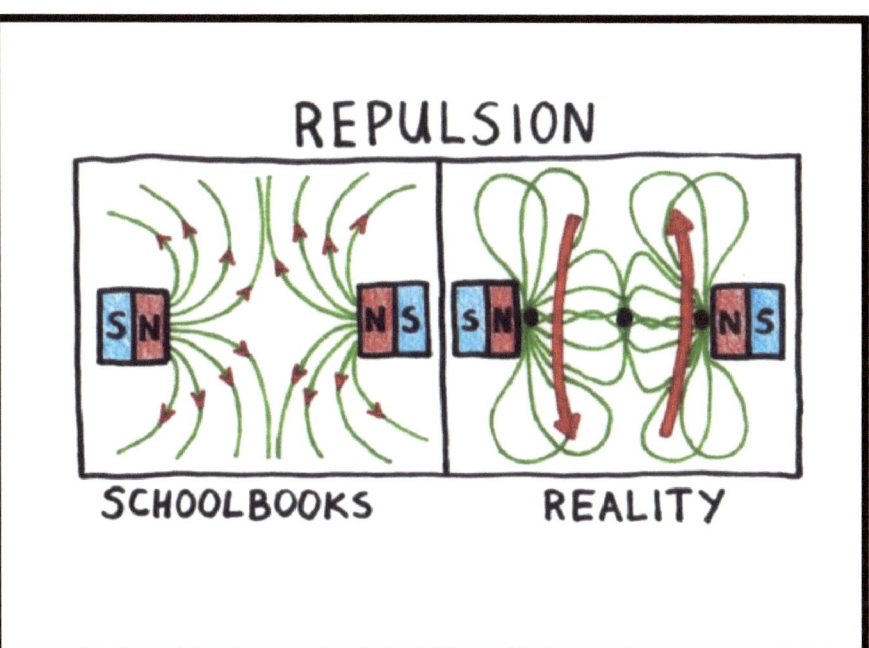
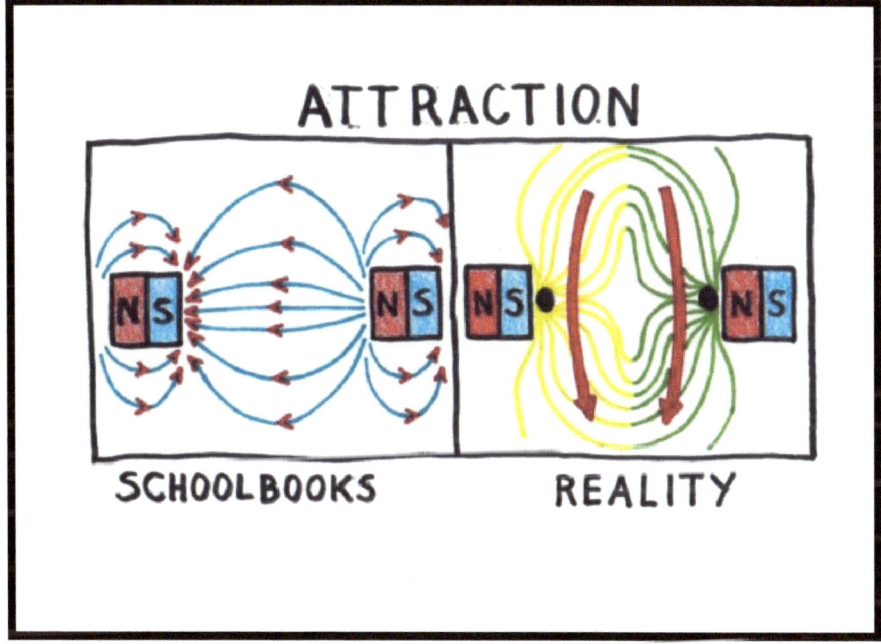

THE TRUE ESSENCE OF VACUUM

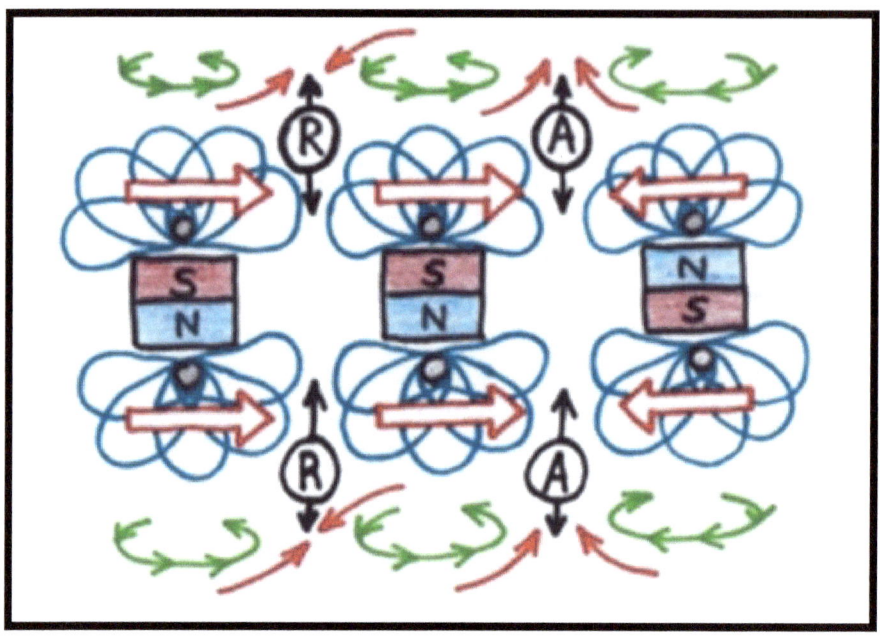

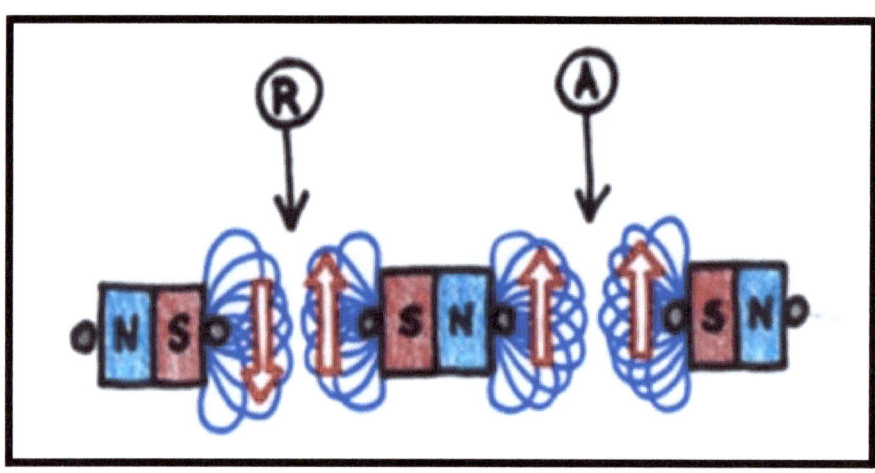

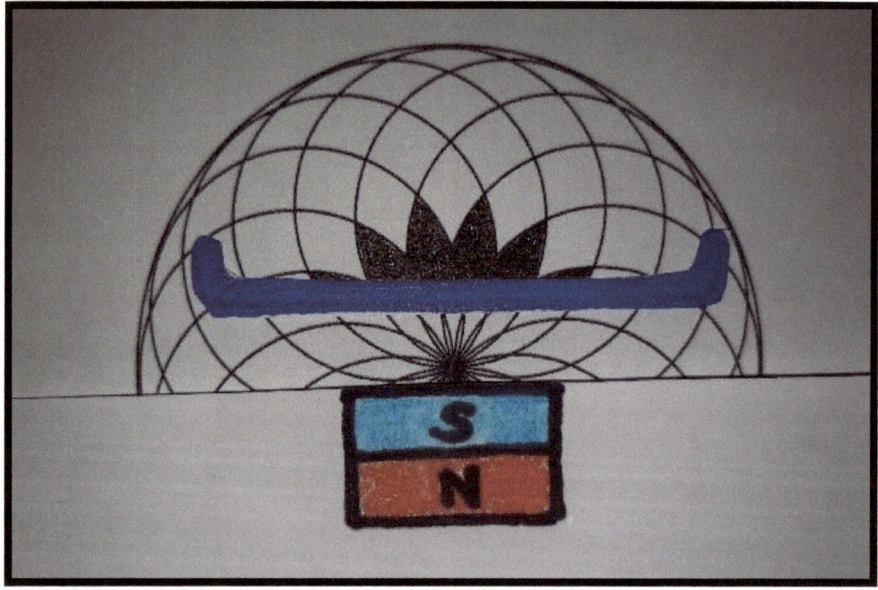

I know why those spikes are being formed! I know how Magnetism and Gravity works! One of my next books will be dedicated to these questions. But you might be interested in one that is already on sale! See the next image!

BOGDANOV

NEW THEORY OF WHERE WE COME FROM AND HOW THE UNIVERSE WORKS: huge blunders of science and abrahamic religions that you did not even know about exposed as never before

Format: Paperback

Price: £20.20 ✓prime

FREE Delivery by **Tuesday, 12 May** Details
Usually dispatched within 8 days.

Thanks for reading! Please add a short review on Amazon and let me know what you thought!

And thanks to Derek Murphy for a free template that helped to make this book more representable.

www.ingramcontent.com/pod-product-compliance
Lightning Source LLC
Chambersburg PA
CBHW040222220526
45473CB00001B/90